GARNISS CURTIS, a pioneer of modern techniques for dating ancient rocks, is founder of the Geochronology Group at the University of California, Berkeley.

CARL SWISHER is a palaeontologist, geochronologist and a member of the Berkeley Geochronology Group.

ROGER LEWIN is the acclaimed author of sixteen popular science books, including four with anthropologist Richard Leakey. He lives in Cambridge, Massachusetts.

JAVA MAN

ALSO BY ROGER LEWIN

Bones of Contention

The Origin of Modern Humans

BY ROGER LEWIN WITH RICHARD LEAKEY

Origins

Origins Reconsidered

The Sixth Extinction

JAVA MAN

How Two Geologists' Dramatic Discoveries Changed Our Understanding of the Evolutionary Path to Modern Humans

Carl C. Swisher III,
Garniss H. Curtis,
Roger Lewin

An *Abacus* Book

First published in the United States in 2000
by Scribner, a division of Simon & Schuster, Inc.

First published in Great Britain in 2001
by Little, Brown and Company
This edition published by *Abacus* in 2002

A CIP catalogue record for this book
is available from the British Library.

ISBN 0 349 11473 0

Printed and bound in Great Britain
by Clays Ltd, St Ives plc

Abacus
An imprint of
Time Warner Books UK
Brettenham House
Lancaster Place
London WC2E 7EN

www.TimeWarnerBooks.co.uk

For our friends and colleagues
who devote their professional lives
to discovering how we, *Homo sapiens,*
came to be who we are.

Contents

Introduction

Three separate journeys converge in this book. Garniss's began in the 1950's, when, with his colleague Jack Evernden, he pioneered the application of radiometric dating to anthropology. Carl got into the same scientific arena by way of establishing that the demise of the dinosaurs was indeed the result of cataclysmic asteroid impact. Meanwhile, Roger has been a spectator and writer of anthropological adventures since the mid-1970s.

Anyone who has been following the progress of anthropology through the years will be well aware of how rapidly the field has been changing, the result of new fossil discoveries and new interpretations. We are honored to be associated with the most recent revolution in anthropological understanding: that of when our ancestors first expanded their range beyond Africa, the cradle of humankind, and what this means for what kind of creature we are. And to underscore the pace of discovery, in May 2000 two Georgian scientists announced the find of two ancient skulls in their country that strongly support the conclusions we reach in our book.

The story we tell has many elements, including history,

science, and the politics of science. For Garniss and Carl, there were many dramas: dramas of discovery; dramas of their colleagues' reactions to their ideas; dramas of simply trying to be able to do their work. Ultimately they were able to do most of what they wanted, and in this they enjoyed the support of colleagues in the United States, Canada, and Java. In particular, Ann and Gordon Getty have been stalwart supporters.

Any book with multiple authors faces the issue of "voice." Although the book was written by Roger, we decided that the narrative element would be best presented in one voice, Carl's. We hope readers will enjoy the story told here, and we encourage your feedback, which can be channeled via Roger's web site, www.TheSoulatWork.com.

CARL C. SWISHER III
GARNISS H. CURTIS
ROGER LEWIN
June 2000

I

Tales a Child Can Tell

"GARNISS, lend me your knife for a second, will you," I whispered. Like most geologists, Garniss always carries some kind of implement for prying small rock samples loose when the need unexpectedly arises, as it so often does. Geologists are as inquisitive as kids, poking into everything. In Garniss's case the implement was a small Swiss Army knife, secured on a cord around his neck, together with a small hand lens. Garniss handed over the knife, wondering what I had in mind.

It was a Monday morning in early September of 1992, and we were in a small, hospital-green room at Gadjah Mada University on the Indonesian island of Java. Aside from Garniss and me, there were half a dozen people in the room. These included the eminent Indonesian anthropologist Teuku Jacob. In his late sixties and recently retired as chairman of his department, Jacob nevertheless retained preeminence in Javanese anthropology. Jacob's longtime assistant, Agus Suprijo, was also present, as were Ann Getty, Meg Starr, and Sharie Shute, colleagues of Garniss's and mine. Conversation rippled easily around the room, but with overtones of excitement and an odd reverence.

A plain wooden table occupied the center of the room, supporting an angle-poise lamp, its connection cord hanging free. On the otherwise plain walls hung an abstract painting titled "The Spirit of Mojokerto," incongruous in its modernity. Cushioned on a thick mat of foam rubber on the table rested a collection of dark brown objects, gnarled and cheating easy identification to the untutored eye. They were ancient human fossils—jaws and teeth, and a skull—which Jacob had just retrieved from the two refrigerator-sized safes that flanked the armored grille entrance to the room. These fossils, the prized objects of Jacob's collection, are rarely seen, even by professionals in the fossil-hunting business. Scholars with serious research programs have to apply to Jacob for permission even to see them, let alone touch them, for scientific study. And even those few who succeed in obtaining official permission have to wait for Jacob's final OK, for he alone is permitted to remove the fossils from the safes. Even Agus, his assistant, is not given access to the vault without Jacob, and the possibility that Agus might study the fossils on his own is out of the question, despite his credentials as an anatomist. Jacob maintains an assiduous—some might say obsessive—protection of the fossils.

Some of the relics arrayed on the table Jacob himself had recovered from the island's ancient sediments; others were found by earlier workers, long ago, going back to the 1930s. Java holds a special place in the annals of anthropology as the source of some of the earliest-discovered fossils of human ancestors. Principal among these is the tiny cranium of the so-called Mojokerto child, who died when he or she was a tender five to six years of age—an event that, by most estimates at the time of our gathering in the hospital-green room, was a million or so years back in human prehistory.

Discovered some six decades ago, the child's cranium has no face and no jaw; and yet, bereft as it is of any recognizable physiognomy, the small, rounded object still elicits a sense of awe in the observer, a connection with a distant past, not so much forgotten as unknown, perhaps even unknowable. Frustrating though they are in their muteness, relics such as these have drawn anthropologists into the search for human origins for more than a century. Their aim is at once the scholarly pursuit of uncovering our species' evolutionary history and a personal mission of understanding what made us human. Hold a fossil like Mojokerto in your hand, and you have what is at the same time the object of scientific investigation into the byways of human prehistory and a petrified moment of your own ancestry.

Garniss and Jacob, friends and colleagues for more than thirty years, were talking about the visits to various fossil sites that we had all made during the previous week and swapping stories about their earlier experiences in these places. Agus remained quiet, out of Javanese respect for his mentor. Ann and Meg were gingerly handling a distorted fragment of lower jaw, feeling the weight of bone turned to rock. Sharie was taking photographs. I was sitting at the table with the child's cranium in front of me. The fossil bone was a rich brown color, the rock matrix black. I slowly picked up the cranium, turned it over, and scrutinized the topography of the underside, thinking about what I was going to do. It would mean big trouble, but if we wanted to make sense of what we had seen at the Mojokerto site a few days earlier, I was just going to have to risk it.

On our visit to the site of the skull's discovery we had seen white pumiceous rock, and yet the rock matrix infilling the child's skull in front of me was black, at least on the surface. If the matrix was black throughout the cranium, it would hold the awful implication that the skull had not been found at the site where we, and the anthropological world, had been told it was found. We would therefore be wasting our time dating the pumice from the Mojokerto section, and what most anthropologists believed—that no one really knew where the skull was found—would turn out to be true. If you do not know where a particular fossil was found, if you don't know which ancient sediments it was buried in, then you cannot know how old it is, because it is from the sediments, not from the fossil itself, that an age can be determined. In that case, no matter how exquisite the fossil, no matter how interesting its anatomy, its value in helping anthropologists piece together the path of human history is much diminished. The child's skull hung in that limbo.

I focused my attention on a large oval lump in the bottom surface of the cranium. Given the cranium's shape, I knew the lump could not be fossilized bone but had to be rock matrix filling the space where the child's brain once was. It would be a good spot.

"Garniss, lend me your knife for a second, will you . . . ?" After Garniss handed it over, I quickly unfolded the blade and started to scrape at the surface of the small raised lump. A fine black dust sprayed out as I worked, moving quickly so that I could get done what I needed to do before the inevitable happened. Just being allowed to see and handle the fossil had been privilege enough. Taking a steel blade to so valuable and valued a relic was sacrilege.

A thundering silence engulfed the room. Seconds passed like hours.

"Garniss," Jacob eventually said, maintaining a tense calm, "come with me. I'd like to talk to you." The two men left the room, and Jacob suggested that they go to his office. It was clear to Garniss, as he told me later, that Jacob was upset by what I had done, but, in a manner that was very typical of Jacob, he spoke politely about various unrelated matters for some five minutes. At last, in a very low-key way, he came to the subject of his concern. "You and I have been friends for many years, Garniss," he said. "And you know that I have to get permission from the committee even to allow you to look at the fossils, let alone do anything with them." Garniss thanked Jacob for his help with the project and for making it possible for us to have access to the fossils.

That was all that was said. Beneath the pleasant exchange, though, there was considerable tension. "I knew that Jacob himself was the committee, that he didn't have to seek permission from anyone," recalls Garniss now. "He didn't mention the knife, and neither did I. But by the way he spoke, and in what he didn't say as much as in what he did, he made it very clear that he thought we had overstepped an important barrier."

It was only when Jacob had left the room to talk with Garniss that I felt free to display my excitement at what I'd seen.

Separate Journey, the Same Goal

We had flown into Yogyakarta a week earlier, on 30 August, after attending the 29th International Geological Congress in Kyoto, Japan, where we participated in a session on geochronology, the science of determining the age of rocks. This branch of geology is extremely important when you are reconstructing Earth history, including the story of human evolution as it was played out in places such as Java. But Java wasn't much in my mind at the conference. In fact, I hadn't given Java much thought at all while planning this trip to the South Pacific, because I had dinosaurs on my mind, not human ancestors.

Ever since the maverick Nobel Prize-winning physicist Luis Alvarez had shaken up the world of paleontology with his suggestion in 1980 that the Age of Dinosaurs was brought to an abrupt end 65 million years ago when a giant asteroid slammed into the Earth, dinosaurs were on the minds of many people whose business it is to find the age of rocks. Alvarez's claim was based on the discovery of unusually high levels of the element iridium, together with the unique ratios of it with the other platinum elements, at the transition between the end of the Age of Dinosaurs and the beginning of the Age of Mammals, the so-called Cretaceous/Tertiary boundary. Iridium is rare in the Earth's continental rock but common in asteroid rock. The humongous explosion generated by the asteroid's impact—the energy equivalent of a billion nuclear bombs—would have vaporized the asteroid, created an immense crater, and filled the atmosphere with debris that would have darkened the skies for months. Much of life cannot survive for long in such a world; pretty soon, argued Alvarez and his colleagues, half the world's species succumbed, the dinosaurs being the most prominent victims. Eventually, the dust debris settled, creating a clear chemical signal—the dust layer rich in iridium—of a cataclysmic episode in Earth history. Or so Alvarez said.

Profound skepticism is not too strong a phrase to characterize the scientific world's response to the asteroid-impact theory at the time. After all, geologists were deeply wedded to the notion that the Earth's topography on all scales—from the smallest curve in the course of a stream to the highest peak in a large mountain range—was formed by the gradual accumulation of small changes. This perspective goes by the term *Uniformitarianism,* and was developed by a nineteenth-century Scottish geologist and contemporary of Charles Darwin's, Charles Lyell. In developing the uniformitarian view, Lyell overthrew the long-established notion that the Earth had been periodically roiled by great catastrophes, such as global flood. The intellectual brainchild of the French geologist Baron

Georges Cuvier, this view, known for obvious reasons as *Catastrophism,* came to be viewed with deep suspicion because, not unnaturally for its time, the theory had distinctly religious overtones to it. One of Cuvier's proposed thirty or so catastrophes that mark Earth history as written in deep geological strata was said to have been the Noachian Flood. The triumph of Uniformitarianism over Catastrophism in the mid-nineteenth century was therefore seen as the triumph of modern science over anachronistic fancy.

When Alvarez started talking about rocks falling out of the sky as playing a major role in shaping Earth history, he seemed to be trying to drag geology back to the dark ages of Catastrophism. He was, of course, but in a different guise. Almost a decade was to pass before Alvarez's theory met with acceptance, the result of the inexorable accumulation of many different lines of evidence. These days, asteroid impacts are recognized as important in shaping the flow of life during Earth history, possibly bombarding the Earth every 30 million years or so, each time wiping out a large proportion of living creatures in the process. The potential hazard of asteroid impact is now taken so seriously that U.S. government agencies are planning to spend hundreds of millions of dollars to try to detect the next asteroid with Earth's number on it, attempting to keep humans from going the way of the dinosaurs. (Hollywood responded by releasing two major movies on the topic in 1998, *Deep Impact* and *Armageddon.*) One piece of evidence about the putative Cretaceous/Tertiary impact remained elusive even after geologists accepted the theory, however: where was the crater of the right size and the right age?

In the spring of 1992, Luis Alvarez's son, Walter, and his student Alessandro Montanari approached me with a small piece of rock that they had obtained from a well drilled by Pemex Oil in the Yucatán Peninsula of Mexico. The site was within a large circular structure, known as Chicxulub, that some scientists, such as A. R. Hildebrand, had begun to think was the result of an impact from a large extra-terrestrial bolide. Not only was Walter convinced that the rock they held represented the melt rock from this impact, but, given the geology of the site, he argued that the impact was of the right age for the Cretaceous/Tertiary boundary, and might represent the "smoking gun" that they had been looking for.

At that time I was working in eastern Montana, with Lowell Dingus and Bill Clemens, on dating the extinction of the dinosaurs. While I liked the idea of obtaining a date for the melt rock from Chicxulub, I found it difficult to see the association between the dinosaurs of Montana and an impact site in Mexico. Since I had just obtained excellent dates of around 65 million years ago for the last of the dinosaurs in Montana, my curiosity over this impact site led to my collaboration with Walter and Alessandro, if only perhaps to show them that the ages of the two events were different. Much to my surprise and Walter's delight, this turned out not to be the case. Instead, the ages for the two sites were indistinguishable from each other.

When these results were published in the journal *Science* on August 14, 1992, they caused a tremendous sensation in the popular media, which was no surprise. What *was* surprising was that they provoked little negative response from researchers who had previously been loud in their criticism of the impact theory of dinosaur extinction. The intellectual tide had shifted to an acceptance of an impact at or near the Cretaceous/Tertiary boundary. But some skeptics continued to say, "OK, so there was an impact . . . but did it cause the demise of the dinosaurs?" What more of a smoking gun could these people demand than irrefutable evidence of catastrophic impact right on the cusp of the extinction, 64.98 million years ago?

The Kyoto conference was therefore a rewarding time for

me. I was in my mid-thirties, confident, even a little brash, and I loved the notoriety the work generated. Much of my work, like that of most scientists, had gone along unnoticed by anyone but a few specialists in the field, with little public fanfare. The glamor of the media attention over the dating of the Chicxulub crater seemed to me to be my proverbial fifteen minutes of fame. Having never stepped into the arena of human origins research, I had no idea of the potential for notoriety that lurked there, too. Of all the disciplines in science, paleoanthropology has an apparently endless capacity for stirring emotions, in its practitioners and in spectators alike. And of all the disciplines in science, paleoanthropology boasts perhaps the largest share of egos, often engaged in intemperate defense of cherished hypotheses and sometimes in fierce public and private battle with other egos. For me, the journey to Java was to be a journey to a different kind of science. Garniss was not so naive.

Garniss, now approaching eighty-one, has been closely associated with many of the important developments in geochronology, particularly dating volcanic minerals based on the slow conversion of an isotope of potassium to an isotope of argon. The accumulation of the argon isotope acts like a clock, measuring the passage of time: the more of that particular isotope of argon there is in a rock, the older the rock is. Given certain assumptions and calculations, a very reliable age can be worked out, as will be described in a little more detail later in the chapter. Known prosaically as potassium/argon dating, the technique became the pillar of establishing the ages of early human fossils. One of the first occasions on which the technique was applied in African prehistory was to try to determine the age of the first human fossil discovered at the now famous Olduvai Gorge, in Tanzania, in 1959.

For almost three decades Louis and Mary Leakey had scoured

the ancient sediments of the gorge, looking for relics of our ancestors. Signs of prehistoric daily life were everywhere, in the form of hundreds and thousands of simple stone tools, and collections of bones that had once been dinner for the gorge's inhabitants. Then, in a story many times told, Mary Leakey spotted massive, humanlike teeth eroding out of a small slope one afternoon in August 1959. When alerted by Mary, Louis, who had been resting back at camp with a high fever, quickly raced to the spot. The teeth proved to be the tip of a paleontological iceberg, which was the complete (but shattered) cranium of what Louis later named *Zinjanthropus boisei,* or East African Man. The fossil was also nicknamed Nutcracker Man, in recognition of its millstone-like molars.

Louis quickly recruited Garniss (and his colleague Jack Evernden) to find an accurate date for *Zinjanthropus.* The principle of finding the age of an ancient fossil is fairly simple but indirect, since at present there is no reliable way of discovering the age from the fossil itself. If you know the age of the layer of rock that covers a fossil, then you know that the fossil is at least this old. For instance, imagine that an individual died a million years ago, at which point its bones began to be covered by sediments. Imagine also that at some time afterward a volcano erupted, spewing volcanic ash over the local countryside and further covering the bones. If, in the present day, you take a sample of that ash and subject it to dating analysis, you will get a result that is a little less than one million years of age. You can then say that the individual lived somewhat earlier than that date. If you date ash layers above and below a fossil, you can say that the individual lived between these two dates, making the estimate more precise. And if the fossil was actually buried in the ash layer, in an ancient version of what happened at Pompeii, you can be most specific of all: the age of the ash is the date the individual died. Unless, of course, an individual's bones somehow become reworked into ancient sediments that were deposited long before the individual lived, which, via certain tricks of nature, can happen. The geologist's job is to look for clues to the true history of the bones.

Volcanoes have been extremely active in the history of East Africa, which has given paleoanthropologists the opportunity to create a very good time frame for sorting the events of human evolutionary history as they unfolded there. Garniss's landmark application of potassium/argon dating in 1961 produced a shock to the world of anthropology: *Zinjanthropus* had lived 1.75 million years ago (now, with a change in the decay constants of the potassium isotope, the age is recalculated to 1.85 million years), more than three times deeper into the past than anyone had imagined. The discovery of the fossil, and its startlingly ancient age, propelled Louis Leakey to world fame. The collaboration between Garniss and Louis seemed set to ripen into a productive partnership, one that would be of great benefit to the science. Within a few years, however, the relationship soured, as Louis repeatedly refused to believe dates that Garniss and Evernden produced for other fossils, when the dates did not jibe with what Louis wanted or believed. Garniss developed a tremendous respect for Louis Leakey the man, but not for Louis Leakey the scientist. Eventually Garniss vowed never to set foot in East Africa while Louis was alive. He kept his vow, although not for as long a time as he was prepared to; Louis Leakey died in October 1972.

About a decade after the breakup with Louis, Garniss was locking horns with another Leakey: this time it was Richard, Louis and Mary's equally famous son. Shortly before Louis's death, Richard found a large-brained human ancestor in the sediments on the eastern shore of Lake Turkana, in northern Kenya. Richard was convinced that the skull, known simply by its museum accession number of 1470, was the earliest known specimen of our own lineage, that is, *Homo.* Using an advanced version of the potassium/argon dating technique, British geochronologists claimed that 1470 had lived and died 2.6 million years ago, making it more than half a million years older than any previously discovered member of the genus *Homo.* The skull did for Richard what *Zinjanthropus* had done for Louis: it propelled him to worldwide recognition.

Very soon, however, a swell of evidence began to imply that 1470 was in fact less than 2 million years old, not 2.6 million. Richard would hear none of it, declaring that the science used to date the skull was the most modern available, and therefore the age must be right. Besides, being more than half a million years younger than the declared age would diminish 1470's importance in the world of anthropology. A tremendous fracas boiled up; accusations and counteraccusations of perfidy—and worse—were hurled around. Garniss, having been sucked into the fight, eventually put the nail in 1470's coffin: he showed that the skull was a little less than 1.9 million years old. (An Australian geochronologist, Ian McDougall, produced a similar age for 1470 at about the same time as Garniss's work.) Only recently has Richard quelled the anger he felt over the episode.

So, Garniss was not naive about what it means to be a journeyman geochronologist in the world of paleoanthropology. He had no illusions about what going to Java might mean, and, in any case, he had had a preview: our August 1992 trip to Java was not his first. He had gone there in 1969, fueled by National Science Foundation money and with the hope of applying potassium/argon dating to the Mojokerto child. The age of the Javan fossils is important to anthropologists because it relates to when humans first expanded their territory beyond the African continent. The human family originated in Africa, just as Charles Darwin predicted over a century ago; no self-respecting anthropologist doubts that, and we now know our prehistory stretches back about 5 or 6 million years. But when did human feet first touch Eurasian soil? There was (and still is, in some people's minds) a lot of doubt about *that.* The Mojokerto child might hold the answer—if an accurate age could be established, that is.

The answer Garniss got in 1969 when, back in Berkeley, he analyzed some volcanic rock samples he had taken from the Mojokerto fossil site was 1.9 million years. This was almost twice as old as prevailing anthropological theory allowed, so the result was sensational. Or rather, unbelievable. "Everyone said I must be crazy," Garniss now recalls. "They said, 'You must have got it wrong,' and that was that." There is nothing more intransigent than a group of scientists clinging to a cherished theory in the face of counterevidence. There was room for some doubt about the age Garniss had obtained, however, because the volcanic rock he had used for the analysis was less than ideal. "It contained very low levels of potassium, and that compromises the accuracy of the date," Garniss concedes. "Low levels of potassium means that low levels of argon are produced in the rock, through radioactive decay. And with low levels of the two elements to measure, there is a significant margin of error in the result, with the technology available at the time." In this case, Garniss calculated the age of the Mojokerto child to be 1.9 million years, but with a margin of error of plus or minus 25 percent. This meant that the child might have lived as long ago as 2.5 million years, or as recently as 1.5 million. "Even the young age is still a lot older than everybody believed," Garniss points out. "But still people refused to listen. They had their idea of what the date *should* be—that is, less than one million years—and that was that. No discussion."

It has to be admitted, however, that the way the date came into the public arena didn't exactly inspire confidence in Garniss's claim. For instance, the initial announcement of the work was contained in two short paragraphs reporting on a symposium at the University of California in June 1970. The date of 1.9 million years was erroneously said to apply to an entirely different fossil specimen from the Mojokerto child, a mistake that was corrected in the pages of the journal *Science* not long after. Even with the error corrected, Garniss admitted in several publications that the low level of potassium in the mineral and the high level of atmospheric argon "greatly reduce [the date's] precision and accuracy."[1] Now Garniss simply says, "It was the best I could do at the time."

Two decades later, Garniss was prepared to try again. The reason? As so often happens in science, new technology made possible a greater precision. The potassium/argon dating technique had developed to the point at which it was possible to work with extremely small samples, rather than on the samples of several grams that were required previously, and with minerals low in potassium, as in Java. This allowed for much more precise dating of rock minerals there.

A New Era of Geochronology

The modern approach for dating rocks goes back to the beginning of this century, when Ernest Rutherford suggested that the natural radiation in them might be exploited as built-in clocks. With information about the rate of decay of certain isotopes and a way of measuring the decay products, rocks may be dated, he said. There are different kinds of clocks, based on different isotopes. Since the 1950s three principal clocks have been exploited for rocks: uranium series, rubidium/strontium, and potassium/argon. This last system, or at least a version of it, is the most useful technique for anthropologists interested in early human prehistory because it works in much of the relevant time range and is applied to volcanic rocks and minerals.

Many rocks contain traces of potassium, particularly some minerals in volcanic rocks, such as potassium feldspar, amphiboles, and various micas, common minerals of volcanic eruptions. Naturally occurring potassium contains a small quantity of potassium-40, an isotope that decays slowly and at a known rate to produce argon-40, a noble gas. During volcanic eruptions the hot molten lava, rising from a great depth in the Earth, loses whatever argon it contains as the pressure lowers and the lava erupts onto the surface of the earth. This effectively sets the argon clock to zero. As time passes after being expelled by an eruption, a potassium-containing rock will accumulate higher and higher levels of radiogenic argon-40, the product of decay of potassium-40 in the rock. Devised in 1948 by Tom Aldrich and Alfred Nier of the University of Minnesota and developed by the Berkeley geophysicist John Reynolds in the 1950s, the potassium/argon clock operates on the simple principle that the more radiogenic argon-40 a rock contains, the older it is. The amount of potassium in the rock or mineral sample has to be measured, of course, because the clock is based on the quantity of argon-40 that would accumulate over particular periods of time from a potassium source of a certain amount. Then, with appropriate technical adjustments, the potassium/argon clock can give reliable dates, particularly for relatively young rocks. By "young" we geologists mean a million years or less rather than hundreds of millions of years. (The technique is also suitable for rocks as old as several billion years.)

To date a rock using the conventional potassium/argon technique therefore requires two separate measurements: the amount of potassium in the rock or mineral, and the amount of radiogenic argon-40 that has accumulated in it. The two measurements involve two separate experiments on two separate samples of the rock, using different methods, since argon is a gas and potassium is a solid. It is a cumbersome but workable approach, one that has been the workhorse of geochronology for many years. But in 1965 Craig Merrihue, a graduate student of Reynolds and Garniss at Berkeley, hit upon an idea that would allow the two measurements—of potassium-40 and argon-40—to be made simultaneously on one sample. Naturally occurring potassium contains a large and constant proportion of a second isotope, potassium-39, which under normal circumstances is stable. Blast it with neutrons in a reactor, however, and it becomes argon-39, cousin of argon-40. Merrihue realized that argon-39 could therefore serve as a vicarious measure of potassium-40 in a rock. When the irradiated minerals are fused at high temperature within vacuum, both forms of argon gas are released and can be determined almost simultaneously in a mass spectrometer: the argon-39 provides a measure of the potassium contained in the sample, and the argon-40 represents the cumulative ticks of the radiogenic clock. All the information required for dating the sample is therefore obtained from the same sample, at the same time, using the same method.

Merrihue developed this so-called argon-40/argon-39 technique with Grenville Turner, who until recently was at Sheffield University, but Merrihue never saw it applied, as he lost his life in a climbing accident in 1966. Various labs were involved in refining the new technique over the next decade, including those of Turner in Sheffield, Jack Miller in Cambridge, Brent Dalrymple and Marvin Lanphere in Menlo Park, and Derek York in Toronto.

Fusing a rock sample in the argon-40/argon-39 technique usually requires raising its temperature to about 1600° C, which was achieved in the early years through radio-frequency induction. The sample was enclosed in a crucible at high vacuum and heated for some 45 minutes. The released gases were first passed through scrubbers to remove reactive ingredients and then expanded into a mass spectrometer, a machine for measuring the amounts of argon. It was a time-consuming business; one sample might occupy a machine for an entire day. The recent revolution in geochronology—taking the argon-40/argon-39 technique to a new level—has focused on the means of heating the sample and on the possibility of using small samples.

In the late 1960s George Megrue of the Smithsonian Institution in Washington, D.C., was experimenting with ways of analyzing the gas content of minerals. He decided to use a pulsed laser for heating small mineral samples, enabling him to drive off the noble gases and analyze them in a mass spectrometer. Later he heard about Merrihue's argon-40/argon-39 dating method and applied it to his system in dating lunar samples in 1973, effectively substituting the pulsed-laser heat source for the radio-frequency heating. Oliver Schaeffer, of the State University of New York at Stony Brook, then applied the pulsed-laser dating technique in a more extensive lunar dating study to samples from the early Apollo missions, but had hardly gotten started before he died suddenly. Derek York, inspired by Shaeffer, brought the technique into the realm of terrestrial rock dating in the early 1980s, using a continuous laser. Soon after York and his colleagues had their single-crystal system up and running, in the mid-1980s, Garniss visited the Toronto lab and was very impressed by what he saw.

Yet to be achieved, however, was the ability to work with *young* single crystals (that is, just a few million years old). When you have one crystal rather than a thousand, you have only one thousandth of the amount of gas coming off. With older samples, there is still a measurable amount of gas involved. Young crystals, however, need a very sensitive mass spectrometer to measure the tiny amount.

Garniss had stuck very successfully with conventional potassium/argon dating during his long career at the University of

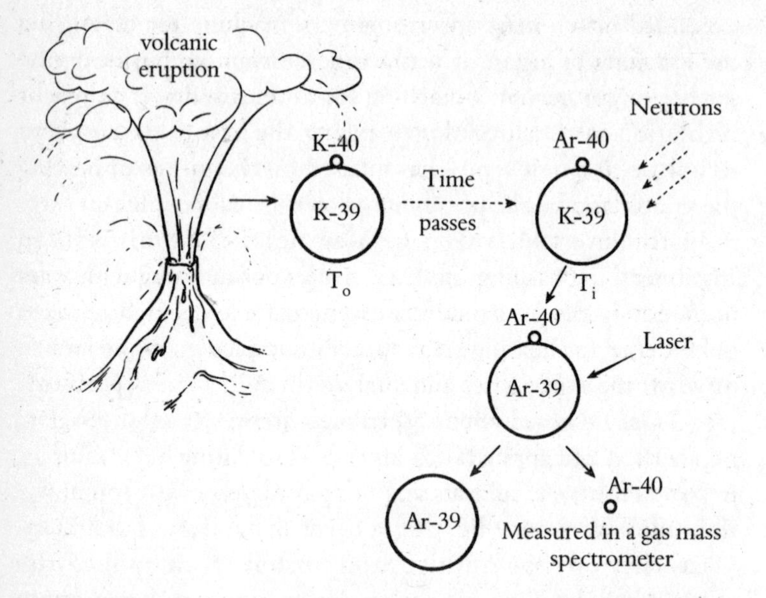

The argon-40/argon-39 dating method: Volcanic ash contains potassium-rich minerals, such as feldspars, amphiboles, and micas. A small percentage of the potassium exists as a radioisotope, potassium-40, which has argon-39 as one of its decay products. In the laboratory, crystals of feldspar are irradiated with neutrons, which converts the stable potassium-39 isotope to argon-40. The crystals can then be individually heated by laser beam, and the emitted argon-39 and argon-40 measured in a gas mass spectrometer. The argon-40 represents a measure of the total amount of potassium that was present in the crystal, and the argon-39 provides a measure of the time since eruption.

California at Berkeley, not venturing into argon-40/argon-39 territory. "But when I saw them measure individual components of gas from a single crystal," Garniss remembers of his visit to York's lab, "I was converted." York's mass spectrometer was not sensitive enough to date young crystals, but new, high-resolution ones were being developed (York had one on order), and it was clear to Garniss that a new era in geochronology was at hand.

At the time Garniss was nearing retirement from Berkeley, although not from geochronology. With his well-equipped, world-renowned lab and a coterie of bright young geochronologists, he was looking for someplace other than the university to set up shop. Donald Johanson, famous for discovering Lucy, the 3-million-year-old human skeleton, in Ethiopia in 1974, had recently established his Institute of Human Origins just up the street in Berkeley. "The institute's board thought it would be a good idea to broaden the institute's scientific base to include geochronology, which is an important part of human origins research, and so they invited us to be part of the institute," Garniss remembers. "It seemed like a perfect marriage, to join Don in his venture."

And so Garniss and his people packed up our lab and moved into the basement of the Church Divinity School of the Pacific, which housed Don's institute. Very soon the lab was making important contributions to the development of single-crystal laser-fusion dating. Primarily through the programming genius of Al Deino and the determination of Brent Turrin, Mac McCrory, and Tim Becker, we were the first to automate the entire process.[2] Within a couple of years we had a system operating that could produce the ages of over 200 crystals in one fully programmed run. Age determinations could now be extremely accurate, even on minerals with low levels of potassium in them, as in Java.

Plans for a Return to Java

"I realized that the argon-40/argon-39 technique could give us a way to settle the dating of the Java fossils accurately, once and for all," Garniss recalls. So in the spring of 1990 we unearthed some of the mineral samples he had collected in Java in 1969, and ran them. The preliminary dates: 1.7 million years, with very little margin of error. That wasn't 1.9, as Garniss had obtained previously, but neither was it the one million years that everyone knew "must" be true. In fact, the slightly younger age we obtained was not surprising, given that the new argon-40/argon-39 system allowed us to date much smaller amounts of material. By hand-picking crystals from the pumice rather than processing bulk samples, we were able in the 1990 test to exclude slightly rounded crystals that might have been reworked from older deposits, thus avoiding parts of the original sample that might have contributed to an erroneously old date. In any case, the 1.7-million-year result was intriguing enough to encourage us to seek funds for collecting more rock samples from Java for a whole new set of dating analyses, which would be done on small populations of single crystals carefully selected to avoid any contamination.

The budget of the Institute of Human Origins, with which our geochronology group was now more formally allied, had a so-called contingency fund, which was meant to support low-budget, pilot projects of the sort we had in mind. A modest figure of $6,000 was agreed upon at a meeting in the fall of 1990 for a new venture to Java. But it wasn't until the late summer of 1992 that the Kyoto geochronology congress provided the opportunity to make use of it. Visiting Java on the way back from Japan would be an inexpensive diversion for us. It would also be eye-opening in our quest to produce accurate ages of these important human relics from Java at last, and, in the process, perhaps dramatically alter our understanding of human prehistory: namely, when did humans first wander out of Africa, and how did we come to be the species we are today?

2

The Road to Trinil

THE sound of *gamelan* music permeates Yogyakarta, the second-largest city on the island of Java. The music flows from unseen sources behind garden walls, fills hotel lobbies, is ever present at weddings, and even bubbles from transistor radios carried aback the ubiquitous *betjaks,* or pedicabs, that swarm the city's choked streets. To Western ears, it is a hard-to-describe, haunting, exotic sound, but Jaap Kunst's evocation, in his *Music of Java,* comes close: "[Gamelan] is comparable to only two things: moonlight and flowing water. It is pure and mysterious like moonlight, and always changing like water." Gamelan *is* Java—the soul of Java—although natives of different geographical regions of this tropical island jealously assert that their version is the superior form. Played mainly on brass percussion instruments, together with a few stringed instruments and flutes, gamelan is said by the Javanese Chronicles to have been introduced to the island half a millennium ago by the nine saints of Islam as they promulgated their religion through Indonesia, west to east. Whatever its true origin, the liquid, bell-like, siren sounds of gamelan signal to the Western visitor that Java is a very special place.

In his *Malay Archipelago,* published in 1869, the British naturalist Alfred Russel Wallace described Java as "probably the very finest and most interesting tropical island in the world."[1] In 1861 Wallace had spent three and a half months in Java, studying its wildlife, people, and history. It was while he was surveying the neighboring island of Ternate, in the Moluccas, a few years earlier that Wallace had come to the same insight about the mechanism of evolution—that of natural selection—that Charles Darwin had been struggling with for decades. The two men published a joint paper on their collective thoughts in 1858, and Darwin then went on to elaborate the idea at great length in *On the Origin of Species* a year later. The archipelago of which Java is a part therefore has a special place in the annals of the study of evolution, dating from the nineteenth century. And before the century was fully out, Java would also find fame in the annals of the study of *human* evolution.

Wallace was entranced by the astonishing variety of life that thrives in such luxuriance in Java, even to the summits of active volcanoes, some of which reach to more than 10,000 feet. Java is green, green, green. "The animal productions, especially the birds and insects, are beautiful and varied," wrote Wallace, "and present many peculiar forms found nowhere else on the globe." Java is also a land of volcanoes, thirty-eight in all, stretched out like vertebrae along the backbone of this 600-mile-long, 100-mile-wide island; it lies almost due east/west, with the Indian Ocean bathing its southern shores and the Java Sea its northern shores. It is a land built by the subterranean stirrings of tectonic plates, which lifted continental rock above sea level not much more than 2 million years ago and are the source of the island's volcanism. Most famous—or infamous—of Vulcan's children in these parts is Krakatau, an uninhabited island some

25 miles off the western tip of Java in the Sunda Strait. In 1883, after centuries of inactivity, Krakatau atomized itself in a gigantic explosion that caused 100-foot tidal waves and claimed the lives of 35,000 people in a brief, catastrophic instant. Reemerging from the waves as it slowly rebuilds itself, Anak Krakatau ("Child of Krakatau") smolders once more.

Wallace also turned his naturalist's interest to the island's recent social history, as evidenced by mysterious, half-hidden ruins of past civilizations. "Scattered through the country . . . are found buried in lofty forests, temples, tombs, and statues of great beauty and grandeur," he wrote, "and the remains of extensive cities, where the tiger, the rhinoceros, and the wild bull now roam undisturbed." These days, sadly, these exotic beasts roam undisturbed no longer, as the booming human population has pushed into virtually every habitable corner of the island, leaving precious little undisturbed nature remaining. Even the steepest of hillsides are terraced, to sustain rice, coffee, teas, and other tropical crops. The eye is enraptured by the smooth angularity and multiple shades of green of these daunting constructions, and the mind is awed: the hillsides are magnificent verdant sculptures. But there has been a terrible price, and that is the loss of much of wild nature. Java is a land of 115 million people in an area the size of California, giving it a population density more than twice that of even Japan.

The modern Javanese are the proud and joyful inheritors of a rich and unique cultural fabric, which has harmoniously woven together the threads of first Buddhism and then Hinduism during the first millennium, and Islam during the second. You can witness this in their daily lives, in their dance, their music, their symbols, and their richly decorated fabrics. A Sanskrit phrase that means "They are many, they are one"—or, as it is more commonly translated, "Unity in Diversity"—is the motto of Indonesia. The Javanese people have less embracing feelings about the Portuguese and Dutch colonists, who, in more recent times, exploited the land and its people and did not become part of their soul. But the accident of history that brought the Dutch to what they came to call the Dutch East Indies in the mid-seventeenth century, thus endowing it with a prominent place on the world's emerging geographical map, also led to its later gaining a prominent place on the world's anthropological map.

An Introduction and a Reacquaintance

We arrived at Yogyakarta's small airport, a short hop from the nation's capital of Jakarta, around noon on August 30, 1992. Sharie Shute was with us; Ann Getty and Meg Starr were to join us a few days later. It was hot and humid, and an aromatic fragrance hung in the thick air. Jacob was there to meet us, as arranged through the exchange of many faxes during the previous few weeks. He and Garniss greeted each other like the old friends they were. I got a more formal handshake and a bow of the head, so typical of Javanese courtesy. Jacob had arranged for us to stay at the Ambarukmo Palace Hotel, an imposing white building from the Dutch colonial era, located on the east fringes of the city. The teeming life of the streets en route to the hotel was no surprise to Garniss, although the population of the city had increased dramatically since he was last here, in 1969. But for me the experience was a shock. It seemed like complete chaos—so many people, so noisy, and the pollution from the traffic! And yet in the distance I could see lush vegetation and streams. It was paradoxical, the chaos and tranquillity side by side.

The foyer of the Ambarukmo Hotel was enormous. Liquid gamelan music swirled around the white arches and through the potted tropical vegetation. After ensuring that we were properly installed in the hotel, Jacob bade us goodbye and arranged a gathering for the following morning, Monday, August 31. "You must be tired from your journey," he said. "Get some rest tonight, and come to my office at ten in the morning; we will look at some fossils." He was wrong about our being tired. Exhausted would have been more accurate. Garniss and I were both suffering from conference fatigue. Even the long hop from Tokyo to Singapore, and from there to Jakarta, hadn't afforded enough sleep time for recovery.

We decided to eat early, and ordered *kangkong* and *burung dara goreng* (which turned out to be spinach and deep-fried pigeon) on the hotel's terrace. We talked idly about unimportant things, preferring instead to listen to the fragments of the Koran that drifted around us, caught on the wind from many parts of the city. We watched as smoke coiled up from the summit of nearby Mount Merapi. As darkness fell, an eerie incandescence developed in the sky above Merapi's caldera, reflecting the glowing debris that from time to time flows down its southwest flank.

"You know, Garniss," I said, "if that thing blows and the lava comes down the southern flank instead, there's nothing to stop it from reaching our hotel!" Merapi, one of the world's most active volcanoes, has erupted with devastating effect more than once in Java's recent history.

"Thanks a lot," Garniss said, laughing. "That sure will make for a peaceful night's sleep!"

A Fleeting Impression

Our plan in Java was to collect samples of volcanic material, or pumice, from half a dozen or more sites where human fossils had been unearthed at different times during the past century, especially the site of the Mojokerto child, so that we could analyze the pumice back in Berkeley with the new, more accurate dating methods. The child's skull, found in 1936, was—and is—one of the jewels of the prized Javan human fossil collection, but it is not the most historic relic. That claim goes to a skull cap and thighbone recovered in 1891 and 1892 from a picturesque site along the Solo River, near the town of Trinil, in central Java.

Storied relics in the annals of anthropology, these Trinil fossils were found as a result of the dogged efforts of a young medical officer in the Dutch East Indies Army, Eugène Dubois. He was a man possessed of an obsession to find what he referred to as "the missing link," a term coined earlier by the German paleontologist Ernst Haeckel. Dubois's fossils came to be called *Pithecanthropus erectus,* or upright ape-man, and at the time of their discovery they were the most primitive form of fossil human known to science. Remains of Neanderthals had been found four decades earlier in Europe, but most anthropologists believed they were a relatively recent racial form of modern humans. In any case, *Pithecanthropus* was undoubtedly older and more primitive in its anatomy than Neanderthal Man. Dubois's discovery eventually prompted a more or less continuous search for further ancient human remains in Java, carried on variously by the Dutch paleontologist Ralph von Koenigswald earlier this century and by Teuku Jacob and various Javanese paleontologists in more recent times.

"There's a simple stone memorial at Trinil that commemorates Dubois's find," Jacob said to me soon after we had arrived in Yogyakarta. "Garniss has seen it before, but it's always worth a second visit." Garniss was only too happy to see the famous Trinil stone again, not least because he was anxious to collect more samples of volcanic minerals for dating, and I was excited at the prospect of seeing some of the historic sites I dimly remembered reading about back in college. "But today you should take it easy, maybe see some fossils, and then visit some sites around the city," continued Jacob affably.

It was Monday, the day after we had arrived in Yogyakarta, and an easy schedule was welcomed by all. Jacob showed us some fossils in the lab at Gadjah Mada University, including the Mojokerto child. This was a week prior to the penknife incident. It was an easy, casual occasion, this first viewing, and the child's cranium was as much an object of curiosity as a target of scientific scrutiny. We were, of course, very much aware of the big question mark that had long lurked over the child's cranium, the uncertainty about where the cranium was unearthed. It doesn't matter how precise your dating technique might be; if you are dating the wrong layer of volcanic ash—one that has nothing to do with when the individual died—then you will get an incorrect age for the fossil. At the time of our trip to Java it was common to hear anthropologists say, "Everyone *knows* that the site of discovery of the child's cranium is just guesswork, so you are never going to get a reliable age. Never."

However, lying unread by most contemporary anthropologists are two scientific papers written half a century ago by researchers who visited the child's site in 1938, two years after the fossil's discovery, taken there by the discoverer himself.[2] They were Helmut de Terra, a Dutch geologist, and Hallum Movius, an archaeologist at Harvard. Both researchers were certain of the location of discovery, and both explicitly stated that the volcanic material packing the cranium matched the color of the pumice in the excavation from which the fossil was recovered. In company with most of our colleagues, however, we had not read these papers at the time of our visit, and so we had not seen this important statement. If we had, we might have devoted a little more scrutiny to the color of the child's skull, and particularly the material filling it, before going to the purported site of its discovery: the fossilized bone, a dark brown, covered with a shiny shellac; the underside, where the volcanic material was exposed, black. But instead of a studied knowledge of the cranium and its color, we took with us into the field only a fleeting impression.

A Tour Through Place and Time

When people think about visiting the sites of discovery of ancient human relics they usually imagine that it requires traveling to remote corners of a tropical country, far away from civilization, with only the sometimes dangerous wildlife of the savannah for company. It's a romantic image, one that appeals to our sense of adventure and sense of mystery about our origins. It also happens to be true—often. Go to the Lake Turkana region of northern Kenya, where Richard Leakey and his colleagues have unearthed so many important fossils, and the romantic image is fulfilled, even surpassed. The same is true of Olduvai Gorge, in Tanzania, and of the Hadar region of Ethiopia, where Don Johanson and his colleagues discovered Lucy and other paleontological treasures. Each of these locations would fit the Hollywood image of adventures in the search for human origins. Not so in Java. In this populous country, you are much more likely to find yourself on the edge of a rice paddy or in the middle of a village, trying to fend off curious children, while collecting rock samples, than in some isolated and arid valley. And journeying between major fossil sites requires driving on Java's busy, narrow highways, not negotiating lava-strewn plains or rutted dirt tracks with dangerous wildlife for company. Facing hungry lions is less hazardous.

"You have to suspend a profound disbelief in your likely survival when you drive on the roads in Java," I told my American friends later. "Most roads have just two lanes, one in each direction, but no one takes much notice of that. If there's more traffic going north than south, then north gets two lanes and south gets out of the way." Buses, trucks, cars, heavily laden bicycles, horse-drawn carts, pedicabs, pedestrians—all swarm the highways in termite-like fashion, often missing each other by inches via heroic last-minute maneuvers. Horrible collisions attest to occasional miscalculation or to sheer stubbornness, when one driver or the other refuses to be the one to give way. The wonder is that there is not more carnage than there is.

So, when Jacob suggested we take a five-day driving tour of the most significant fossil sites on the island—giving us an opportunity to collect rock samples for dating the allegedly oldest human fossils outside of Africa—it was with mixed feelings that we agreed. We needed the samples, and we wanted to collect them ourselves. We didn't really have much choice, but I was worried about it. Garniss was much more sanguine.

"Oh, it's not so bad, you know," he reassured me. "When you get to my age, you don't let little things like getting killed worry you much!"

"That's because you drive just like they do," I told him.

Yogyakarta is located in central Java about 25 miles from the island's southern coast. Jacob's planned journey back through Java's prehistory involved driving first northeast, then east, with the magnificent topography of the Kendeng Hills to the north, and to the south the distant giants of Mounts Lawu, Wilis, Butak, and Bromo, to distract the mind from the prospect of imminent death. The mountains themselves can be a source of death, too, as they are still active, from time to time pouring molten lava and belching hot volcanic ash onto nearby villages. But they also bring life to the island because the volcanic minerals enrich the soil and make Javan agriculture bounteous.

Clinging to the roadside are villages made up of knots of small houses, often with simple bamboo walls and red-tiled roofs that curl upward at the eaves. Graffiti is common, but not of the Western variety. The word *Berjuang*, meaning "Overcome the struggle," is seen on rooftops, often spelled out in flowers.

The letters "PKK" are painted on many roofs, an abbreviation of the name of the island's women's movement. Two fingers with the words "Anak Cukup" unsubtly means that two children are enough; this government program has proved difficult to enforce. Entire villages are trimmed in yellow or red paint, signifying the political party in charge. The Javanese are still struggling with the notion of freedom of expression in a tightly controlled democracy, and it shows.

Jacob's planned tour held close to the course of the Solo River, Java's longest waterway, rising in the harsh limestone hills of central Java and tracing a tortuous 220-mile course before spilling into the Surabaya Strait, off the island's northeast coast. Once an important mode of economic transport, these days the Solo River is more a scenic backdrop, as well as a laundry and bathroom for local people. To Western eyes, the journey offers a kaleidoscope of rural life of an exotic nature. Peasants in straw hats the shape of an overturned shallow funnel stand thigh-deep in irrigation canals that create checkerboard rice paddies, bucketing water to the thirsty crops. Flocks of ducks

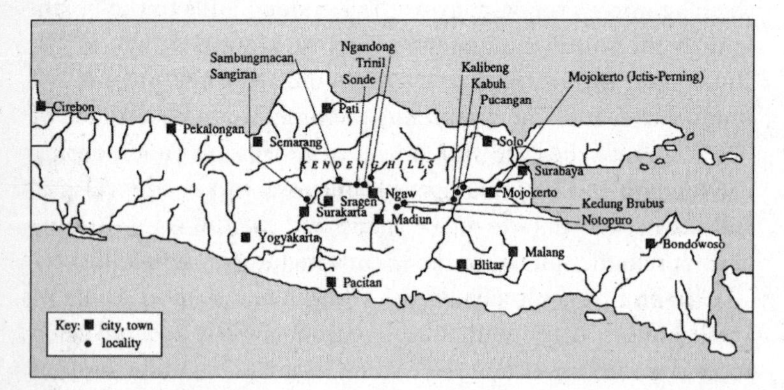

Central and eastern Java, showing the principal sites where early human fossils have been found.

are driven from field to field, providing the dual service of eliminating insect pests and fertilizing the crops. Zebu and Balinese cattle graze lazily here and there while teams of oxen toil in fields, pulling simple wooden plows. Roadside markets offer brightly colored birds and cowering, frightened monkeys for sale. Convoys of schoolgirls, hundreds of them, dressed uniformly in blue Islamic garb, cycle along a country road. And in the distance are the terraced hillsides that are emblematic of Southeast Asia.

The promised trip to Trinil was on Wednesday, three days into Jacob's tour. Such a visit is akin to a pilgrimage to those of us in the human fossil business. We had spent the morning at Sangiran, with Jacob showing us where various specimens had been found, often by farmers while plowing their fields. At Sangiran, as on the day before, we pored over geological maps, trying to match the abstract to the reality. Describing that process as hard would be a gross understatement. For one thing, everywhere we looked there were crops, which obscured the geology. Garniss and I felt as if we were in an *Alice in Wonderland* world where nothing made sense. We had been under the impression that the geology had all been worked out, that all we had to do was collect some volcanics outlined in earlier papers by Japanese and Indonesian scientists, but it now began to dawn on us that we were wrong.

The problem, we would discover, went deeper than our lack of familiarity with the terrain or its being carpeted with obscuring agriculture. The geological maps, created by the Japanese and Indonesian scientists, had a thread of Alice-like fiction to them. They showed no roads or topography, only the general vicinity of small villages and precisely drawn geological contacts that lacked any reference point. The Indonesians leading the trip proceeded without a topographic map of any kind. (In fact, the only available topographic maps were made back in the 1930s, and much had changed since then.) It was like being on a subway in a strange city, stopping at various exits, emerging from underground, and trying to figure out where you were and how this spot related to the last place you had emerged. By the second day we saw that our Javan venture was going to involve a lot more effort than we had expected. As it turned out, it would take five more trips to Sangiran, and many days of working out key geological beds with other geologists, before we finally felt confident that we had a good handle on the geology of the region.

On that first confusing morning at Sangiran, we spent part of our time wandering along muddy streambeds, trying to trace the strata they exposed. At one point, after a long, muddy trek along the course of a streambed, we ended up by the side of a canal that, much to our amazement, was the site we had driven directly to the previous day! Although the precise reason for our muddy excursion was obscure to us, getting to the canal was important, because in the 1980s the excavation of the canal had by chance unearthed an important specimen. This was the front and the back of an ancient human skull, like the one Dubois had found at Trinil. Getting a reliable age for this specimen was one of our goals. "There we were, standing on the bank of a canal, looking for a suitable volcanic ash horizon from which to take a sample," remembers Garniss. "Luckily, we found one. This was important because the skull fragments, known simply as S27 and S31, were thought to be among the oldest human remains at Sangiran and perhaps in all of Java. Down in the canal were a couple of women doing their laundry and looking up at this bunch of guys who were scrubbing around in the dirt like little kids. They must have thought we were crazy." After the volcanic ash specimens were safely stashed, it was off to Trinil, via the district town of Ngawi.

One of the roads from Ngawi leads to a pool containing

sacred turtles. In the center of the town is a square with two old *waringin* trees. These many-branched, many-rooted trees are considered holy by the Javanese, and no one would dare cut one down. Jacob selected a simple restaurant for lunch. The meal consisted of fried octopus, coconut milk with brown sugar and ginger, jackfruit, which looks like strips of beef when cooked, and buffalo skin, which tastes much as one might imagine. Jacob amused himself by watching the distinct lack of enthusiasm of the non-Javanese to eating this Javanese specialty. Except for Garniss, who loved it.

Trinil, just ten minutes from Ngawi, is reached along a narrow country road and then an even narrower dirt track. A crude wooden sign says "Pithecanthropus ahead," and an arrow shows the direction. The road cuts through a flat river plain where rice fields stretch on as far as one's eyes can see. To the north, rows of women, seemingly unaware of our visit and perhaps even of Dubois's history there, were bent sharply at the waist, standing calf deep in mud, placing rice seedlings, one by one, in the mud, all aligned in neatly arranged rows. The rows were interrupted only by a small levee that separated that particular field from an adjoining field of identical appearance. On the south side of the road, two oxen were pulling a simple wooden plow; upon it stood a deeply tanned man dressed only in dark knee-length shorts and a conical hat. We passed women and men walking along the road, each carrying a bundle, some packed with rice, others with wooden sticks.

The road ended abruptly on a high bank above the great Solo River, at a place where it takes a leisurely left-handed sweep as you look downstream. Swallows peeled noisy aerobatic arcs in the blue sky. Fields of cassava and corn were cultivated to the water's edge, where a simple canoe was tethered to a stake. A huge kapok tree offered welcome shade. Stands of bamboo rustled in the tropical breeze. The place had not much changed since Dubois's day, except for the absence of forest, the presence of a recently built museum, and the older presence of the Trinil stone, on which is inscribed "P.e. 175 m ONO 1891/92." This cryptic message stands for, in Dutch, "*Pithecanthropus erectus* wurde 175 meters Ost-nord-ost von dieser Stelle gefunden in den Jahren 1891/1892." Translated, it says, "*Pithecanthropus erectus* was found 175 meters east-north-east of this spot in the years 1891/1892." The site's sparsely equipped museum is home to fanciful, full-size models of *Pithecanthropus,* notable for being extremely hairy and sporting alarmingly long fingernails. In the grounds behind the museum, life-size models of giant buffalo, rhino, deer, are caught in petrified stance, a reminder of the wildlife here a million or more years ago. At the entrance to the museum a life-size model of the elephantlike Stegodon greets the occasional visitor.

Standing by the Trinil stone, you can look out across the river and see the spot where Dubois's workers excavated tons of sand and gravel in their search for *Pithecanthropus* a century ago. Most sites of major fossil discoveries have a disappointingly ordinary look to them. This one is no exception: just a bend in a river and, half submerged in slow-moving waters, an incongruously located sand pit. In his book *Meeting Prehistoric Man,* von Koenigswald described the way, six decades ago, he saw this river scene: "Its waters are brown and murky; buffaloes loll in the water, and this Cloaca Maxima of Central Java carries away all the refuse of the numerous villages along its banks. But this does not hinder anyone here from bathing in the river or cleaning his teeth with its water. The women by the bank are standing up to their waists in the water peacefully doing their washing. Nobody here has yet heard of bacilli and similar fiendish inventions of the West."[3] Although there was no laundry under way on this day, the reality of the river and its local people had changed little since von Koenigswald's time.

We stayed on the museum side of the river and soon found a source of pumice that we thought would be suitable for our dating methods. The moment I started picking up samples, I was joined by a small gang of children who for a while found collecting bits of rock and handing them to this strange foreigner more fun than jumping into the river. I had no idea what they thought I was going to do with what they were giving me. Before long, the novelty of collecting stones for strangers wore off and the allure of the brown river prevailed once again.

The water level was low, so we expected to be able to see Dubois's excavation, which is often under water. We waved down a fisherman who was passing in a simple dugout canoe. He took us across the Solo, where the excavations were indeed clearly exposed, its squared-off trenches that had been dug into the bedrock of the quarry to drain water from the excavations looking much like the remains of a past settlement. We collected volcanic material that we would take back with us to Berkeley, and boarded the dugout for the return journey.

Next stop, Mojokerto.

3

On to Mojokerto

THE Mojokerto child takes its name from the town of Mojokerto in eastern Java, which is about 6 miles southwest of where the little fossilized skull was found in 1936. Two million years ago, the Mojokerto region was a vast delta, with many river channels emptying lazily into the sea. It is easy to find beds of fossilized mollusks that attest to the region's ancient marine past. Today, the region is hilly, and is economically important as a source of oil. However, to most Indonesians, the area of the Mojokerto site is a poor agricultural region. Rising out of the richer, flat lands suitable for rice and sugar cane, the comparatively dry Kendeng Hills support large teak plantations as part of the government's feeble effort to restore the vast teak forests that once covered this region but were leveled after the Dutch colonists left.

We were unable to go straight to the Mojokerto site from Trinil, so we had to spend the night of September 3 in the town of Madiun, almost 25 miles to the southeast of Trinil. Because we had fallen behind our self-imposed schedule, we were forced to make the drive in the dark, a memorable experience.

Javanese truck drivers, clinging to the belief that having headlights lit in the dark is bad for the battery, prefer to drive blind for much of the time. When we weren't contemplating the prospect of imminent and violent death, we allowed ourselves the pleasure of smelling the fragrance of incense as we passed through small villages. We could also see and smell aromatic wood being burned in braziers to cook food—skewers of beef and chicken—along the dirt streets in front of small cottages.

Once a sleepy little provincial town on the border between central and east Java, Madiun is now a busy center of commerce. The Dutch built a small railroad trunk line that connected the surrounding fields with Madiun to carry sugar cane to a processing factory. Many of the sugar cane cars now lie abandoned along the highway, as trucks, typically precariously laden with cane, have unfortunately replaced the utility of the railroad, causing further clogging of the two-lane road that connects Ngawi and Madiun.

Jacob had booked everyone into the Merdeka Hotel in Madiun, where he and Garniss had stayed in 1969. "Must have been Jacob's idea of a joke," recalls Garniss. "It was pretty bad back then: a dollar a day for three inedible meals, a mattress made of narrow rods sewn together, a water cistern for washing and toilet, and the mosquitoes—boy, they were bad! Our driver preferred to sleep on the front seat of our car than to suffer the accommodations of the Merdeka!" The place wasn't much better on this second visit of Garniss's, except that it was more expensive. "Hotel" is misleading—the Merdeka struck me as being more like a trucker's motel, where you drive around back and park in front of the door to your room. Despite its appearance and questionable clientele, there were house rules posted in each room—surprisingly, in English. The first rule read: "No members of the opposite sex permitted in the room after 10:00 P.M., and at all other times the door shall remain open."

The following morning we all set out for Mojokerto, driving first through the little Dutch town of Djetis (or Jetis), whose name von Koenigswald gave to the vertebrate fossil fauna that is associated with the Mojokerto child. About a mile to the east, we reached the small village of Perning with its white houses, brilliant flowers, and laughing children, where the road abruptly made a sharp turn to the north. Our vehicle barely fit between two closely spaced buildings with open fronts selling various foodstuffs and assorted household items. The small dirt road was closed off by a white painted beam blocking our path. After scrutinizing the car, a young woman reached out of the nearby store window and released a rope that permitted the cantilevered beam to swing up out of the way. We waved at the growing crowd of children and continued north for a few miles, through fields of rice, sugar cane, and corn.

The flat topography grew hilly as we entered the Kendeng Hills. Passing through the small village of Kepuhklagen, we began to see outcrops of sediments dipping to the south. We had brought a map with us, which had been drawn by the Dutch geologist Duyfjes in the 1930s, and from it we could tell that we were on the south limb of the anticline that forms the backbone of the Kendeng Hills. We soon reached a flat area where the Kali Klagen, a small, ephemeral stream, had carved out a small valley that was cultivated in cane. After a further half mile, the route started to climb once again along a road that had been cut into the hill, and we could see rock layers in the embankments on both sides of the car. The rocks were now dipping to the north. It was clear that we had crossed the axis of the anticline and should be climbing up section, soon reaching the layers where the Mojokerto child was said to have been discovered. As the car reached the top of the hill, the landscape flattened. Jacob asked the driver to pull over and stop. "We're here," he announced.

On the side of the road was an official, although weathered, sign announcing "Pithecanthropus mojokertensis," with an arrow pointing to the east, "200 meters." It seemed straightforward. On the side of the road was a small white house where three older men sat under an oversize roof on a simple wooden bench. To the right, a woman and a man sat on a piece of cloth laid out on the ground. On it was a pile of reddish fruit from which they were breaking off odd, nut-shaped features. "Cashews," said Jacob, by way of succinct explanation.

Unexpected Uncertainty

With Jacob leading the way, we walked down a dirt path that wound among cashew trees and a few jackfruit trees and along the edge of a deep gully. It was hot and dusty, and none of us talked much. The path dropped down, crossed the gully, and curved back on itself on the other side. When we reached the top of the gully we came face to face with a stone memorial that proclaimed the importance of the discovery of the Mojokerto child's skull at a site nearby. In February 1936 Andojo, a trained Indonesian worker for von Koenigswald, had come across the child's skull at a depth of about three feet, quite by chance. Andojo had been excavating fragments of the skull of a Leptobos, a primitive ox whose horns are placed directly over the eyes and not in the more usual position at the rear of the skull, entombed in ancient sandstone sediments. Leptobos fossils were important to von Koenigswald because they were one of his guide fossils of the Djetis fauna, which he considered to be older than the Trinil fauna. This was part of the basis for his believing that the Mojokerto child was of late Pliocene or earliest Pleistocene age, which would make it around 2 million years old. To Andojo, the child's skull was a great bonus, but in fact it was an even bigger find than he appreciated, because he initially thought it was the skull of an orangutan. Together with other fossils, the child's skull was sent to von Koenigswald, who at once recognized its true identity. "As soon as we had unpacked the eagerly awaited fossil, we immediately realized it was another human skull," he later wrote.[1]

"This is the site," announced Jacob after he led us into a protected gully below the monument. Placing his hand near the middle of the small outcrop, he said, "This is where Andojo found the child."

Garniss was perplexed. This was not at all what he was expecting. "Wait a minute, Jake," he protested. "This isn't the place you showed me last time." Was it true, as many suspected and feared, that Jacob wasn't certain exactly where this important fossil had been found?

Back in 1969, Jacob had identified the supposed excavation site as a rectangular pit about three feet deep and measuring four feet by eight feet, some 60 yards from the place we were now. The pit was in the middle of a barren field, out in the open. "I dug around in the pit for a while and found it was just soil," remembers Garniss. "I asked Jacob if he was sure this was the right place, and he assured me it was. I thought to myself, 'This isn't right; this can't be the place.' " Garniss's incredulity was based on the simple fact that you just don't find ancient fossils in soil; you find them in sedimentary rock, such as sandstone. And if you do find them in soil, then they must have been reworked—that is, eroded from their original site of deposition and redeposited somewhere else—in which case there's no way of finding out how old the fossils are. "In any case, I could see that there was no way this pit had been there for more than three decades," recalls Garniss. The landscape changes very quickly in those parts, and the pit Jacob showed Garniss was too fresh to have been dug in 1936. Garniss began to look for a layer of volcanic ash—or "tuff," as it is known in the trade—that would be suitable for dating, and eventually found one exposed by a small creek not far from Jacob's pit. It was a sample from this tuff that produced the universally disbelieved age of 1.9 million years, three decades ago.

"So why have you brought us here, Jake, and not the pit you showed me back in 1969?" Garniss asked Jacob. "I'm confused."

Jacob explained that in 1975, six years after Garniss's first visit, he had met Andojo, the discoverer of the child's skull. "I asked him if he would take me to the Mojokerto site, and he said he would be happy to," explained Jacob. "This is where he brought me." Jacob told his story very casually, even though the implications were significant. We were dealing here with potentially one of the most important of human fossils, one that could alter the way we understand when and how humans first moved out of Africa, but one that most anthropologists refuse to take seriously because of the uncertainty over where it was found. And yet here was Jacob, keeper of the fossil, casually admitting that even *he* had been confused.

"No wonder people are skeptical about whether Jacob or anyone really knows where the skull came from!" I whispered to Garniss when Jacob was no longer in sight.

"It doesn't exactly build confidence, does it?" Garniss replied. "I just hope he's right this time."

Time and Change

To the untutored eye, the physical landscape around us has an air of permanence about it. True, the vegetation may change with the seasons, but the rock under our feet seems unchanging, enduring, unalterable. The world around us simply does not appear to change, or at least not in important, large-scale ways. That, of course, is an illusion, a personal perception born of the tiny window of time that the human life span affords each of us. In reality, our physical world is in constant transformation, mostly extremely gradually, but sometimes catastrophically. Small changes over vast tracts of time accumulate to reconfigure the continents in dramatic ways.

For instance, when you stand at the Mojokerto site you are witness to the island's history, some 2 million and more years of change encrypted in a geological section some 200 yards in thickness. Near the base of the section (the oldest part) lies a distinctive horizon of fossil shells that Dufyjes in 1936 gave the term Mollusk Horizon II, which attests to a time when ocean was here, not land. When this part of Java rose above the waves, the Mojokerto region was coast, as can be seen from claystones that preserve debris of plants that once carpeted the banks of lagoons. Volcanic deposits are common in the lower part of the section, and at one place they reach almost two yards in thickness. The elephantlike Stegodon as well as hippos, rhinos, antelopes, small oxenlike bovids, deer, and pigs used to live in this region; their fossils make up part of von Koenigswald's Djetis Fauna. Most of the fossils are just broken fragments, but partial skulls and jaws of hippos and bovids and a well-preserved Stegodon tusk have been found in this area. During this first brief visit we found only isolated bovid teeth, pieces of horn cores, and limb fragments.

Like that of the other islands of Indonesia, Java's birth from beneath the waves was the result of tectonic activity in the Earth's crust. But the activity was asymmetric in orientation: the western islands of Indonesia appeared before those in the east, and in Java itself, the western part of the island is older than the east—that is, 7 million years as opposed to 2 million. Being an island, Java could become populated only when animals (and to a lesser degree plants) could migrate there from mainland Asia. Although some creatures—particularly small reptiles—have been known to traverse long ocean distances as passengers on natural vegetation rafts, larger animals require solid connection, or at least a narrow, shallow channel, between their homeland and their newfound land. The only time such a connection or channel comes about in an island chain like Indonesia is when sea level drops substantially, as occurs during major glaciations, which suck water up into massive ice sheets. Once sea level drops by 40 yards, much of the island chain of Indonesia becomes a peninsula, virgin territory for multiple migrations.[2]

The first time that this was possible for Java was, apparently, at some point between 2.4 and 3 million years ago, and the migration window remained open until something less than 2 million years ago, after which it reopened sporadically at later peaks of glaciation. Paleontologists judge these dates by the kinds of animals that turn up in the newly occupied land, a kind of evolutionary clock. In the case of Java, the newcomers in western Java included the modern elephant and deer, which had evolved in Asia in that 2.4-to-3-million-years-ago time period. As the island continued to rise above the ocean to the east, descendants of these first immigrants spread eastward also. One of the immigrants to Java was, of course, early humans, who, like the other animals, were simply expanding their range as opportunity arose, rather than embarking on deliberate migration. Exactly when this happened is, of course, the key question in our quest. *Homo erectus* first appeared in Africa close to 2 million years ago. (*Homo erectus* is the name by which *Pithecanthropus erectus* eventually came to be known.) Did populations of this creature move out of Africa and into Asia immediately, or did the move come much later, as most anthropologists believed at the time of our trip to Java? The migration window into Java was open at that earlier point, 2 million or so years ago, so it is possible—despite anthropologists' skepticism—that *Homo erectus* was among the earliest of the island's new occupants and that the Mojokerto child descended from these first migrants.

A Sample Collected, and an Open Question

"This is the site," Jacob repeated as he placed his hand on the surface of the outcrop we were facing, the place Andojo had shown him.

With the confusion resolved about which was the correct discovery site, we began to scrape clean the surface of the outcrop, using hammers and knives to remove the weathered crust. The exposed rock was a tuff layer that appeared to consist almost entirely of pumice and volcanic matrix. The volcanic material had been disgorged from the belly of the nearby Mount Wilis during what had obviously been a major ancient eruption. The sky must have been darkened by the gigantic cloud of ash, which eventually settled to Earth and carpeted a vast region of the island. The ash clogged the many streams that emptied into the shallow marine bay to the north of the Mojokerto site, and formed thick deposits. No one knows how the Mojokerto child died, and there is no way of finding out. But she must have died near the banks of a river, her body perhaps floating in shallow water for a while, before bone separated from bone as her flesh and cartilage decayed. Where the rest of her body is, we don't know; but we do know that her skull became entombed in the layer of volcanic ash where we stood that day, the cavity where her brain once was becoming filled with ash, all now turned to stone.

We spent much of the day, until nightfall, collecting suitable material to take back to Berkeley for dating. We collected some pumice samples, which Garniss packed away in a sample case. The lumps of pumice were white. We were concerned to see whether there might be pumice layers higher in the section than the one the child's skull was said to have come from. We saw none.

With our sample collection at Mojokerto complete, we began our journey back to Yogyakarta, stopping at the village of Ngandong on the way. The diversion to Ngandong is a two-hour nightmare off the main road. The Ngandong road winds through a government teak forest, and although much of the road is paved, the heavy traffic of giant trucks that use the road has left it more a path of potholes than a serviceable road. By the time we reached Ngandong, we had a collective backache, and the inescapable prospect of retracing our steps home was a dreadful thought. We arrived at a small schoolhouse in the village of Ngandong, which turned out to be the location of one of the fossil excavation pits dug in the 1930s.

In 1931 workers with the Dutch Geological Survey discovered a treasure trove of fossils, including human fossils, not far from the village and close to the Solo River. More than twelve human crania and partial crania were discovered, along with over 25,000 fossil vertebrates. Finding so many human fossils is unusual, but these were relatively recent—some 100,000 to 400,000 years old, by most estimates. Whatever their age, the Ngandong skulls were considered some of the most recent examples of *Homo erectus,* principally because although the skulls looked very much like that species, in terms of the shape of the brain case and the thickness of the bone, their brains were larger than is typical. We wanted eventually to try to pin down the age of the Ngandong fossils more precisely, a kind of side project to the more important task of getting a secure date for the Mojokerto child. We considered it a simple, routine task; we had no idea how important—and startling—our results would eventually turn out to be.

We arrived back in Yogyakarta late on Saturday, 4 September, tired and still more than a little disconcerted with what we had learned—and not learned—in the field: the confusion over the geological maps, and the shifting location of the putative site of discovery of the Mojokerto child. We had survived the journey, for which we were thankful. We decided a day of rest was in store.

Ten miles east of Yogyakarta is Prambanan, Java's largest Hindu temple complex. Hewn from coarse, black lava, this group of temples centers on the towering temple of Siva, god of destruction and creation. Every surface of the temple is replete with Hindu iconography, including a detailed depiction of the epic tale *Ramayana,* a complicated intrigue of love and lust, power and loyalty. As we toured this place of thoughtful tranquility, a fierce storm suddenly blew up: high winds hoisted into the air corrugated sheet metal and scaffolding that were part of the continuing restoration, and the sky turned an ominous dark purple-gray. We should have seen it as an augury of things to come, but we didn't. Instead, we simply hurried back to Yogyakarta to make preparations for the following day's more thoughtful and detailed examination of Java's precious collection of human fossils, in particular the Mojokerto child.

A Black and White Problem

When I saw the child's skull that Monday morning, September 7, I thought to myself, "Oh my God, we have a problem here."

For the first time I was looking at the Mojokerto skull with the eyes of a geologist, not those of a passing observer. This time I saw that the bone had been well preserved during the slow fossilization process, which had turned it a rich brown color. The layer of shellac that had been applied to the skull many years before made the color richer still, and shiny. I then turned the skull over so as to examine the volcanic matrix that filled the cranium. It was black, a dull black. "Look at this, Garniss," I said in a quiet aside. He looked at where my finger was resting on the black matrix, and then at me. We both understood the implication of what we saw.

There was nothing black in the rock layers where Jacob had said the child's skull had been found and where we collected pumice samples, which were white. Didn't Jacob understand what this meant? It meant that the skull must have come from somewhere else. It meant that the skeptics might be right: no one knew where the skull had been found.

"We're screwed," I lamented to myself. "Screwed. Now what do we do?" Then I noticed something odd about the black matrix. It was hard to be specific about it, but it just didn't seem right to a geologist's eye.

"Jacob, what's this black stuff?" I asked, indicating the underside of the cranium.

"It's manganese," Jacob replied. "Manganese staining."

Geochemistry is a creative process, so that fossils and other minerals can become strikingly colored through time. One such process, manganese staining, produces a distinct black hue. The color can sometimes lead to erroneous conclusions. For instance, the layers of blackness that were once thought to be the remains of hearths in the famous Peking Man cave, in China, recently turned out to be the result of manganese staining, not evidence of fire. "Manganese staining," I ruminated. "Are you certain?" I asked Jacob. Jacob said he was. He had obviously not recognized the incongruity of the black matrix that filled the skull and the white volcanic ash layer from which he believed the skull had been recovered.

I held the cranium's lower surface close to the light and turned the fossil through several angles so as to assess the color more acutely. "Jacob may be certain, but I'm not," I thought. "It's too shiny for manganese staining. I have no idea what it is—paint, maybe; but I'm just going to have to find out."

It was then that I turned to Garniss and whispered, "Garniss, lend me your knife for a second, will you . . . ?"

My heart leapt when I scraped away the black surface and found white pumice underneath. Until then I had really feared that we'd have to forget the whole Mojokerto project as hopeless. Seeing the white pumice in the child's cranium, just like the pumice we'd seen at the site, made it at least credible, if not certain, that the ash layer we had seen a few days earlier was indeed the genuine site of discovery.

4

The Lure of the Missing Link

On 29 October 1887, Eugène Dubois and his young wife, Anna, boarded the steamship *Prinses Amalia* and sailed from their native Holland to the tropical island of Sumatra, Java's neighbor, in what was then known as the Dutch East Indies and is now part of Indonesia. Little is recorded of Dubois's thoughts and reflections during the six-week voyage, or of his impressions on landing in the seaport town of Padang on Sumatra. But the contrast between the two worlds could hardly have been greater. Amsterdam: urbane, worldly, and sophisticated. Padang: rustic, parochial, and simple. One a city of learning, the other a town of illiteracy. Christian Holland, Muslim Sumatra. Voluntarily abandoning temperate Europe for the steamy tropics required a determination of uncommon proportions a century ago, and determination was a commodity Dubois did not lack. His goal was to find the missing link between apes and humans, a transitional form that would open the book of human prehistory. He gave no heed to the deprivations and hardships of life in the tropics because, in his driving obsession, he felt he knew where to look for his prize, and he had no doubt that he would succeed.

The fact that Dubois—or anyone—would set out to find the fossil remains of a human ancestor a century ago was extraordinary enough, for two reasons. First, although human fossils existed—such as those of Neanderthals—none was widely accepted as being part of human ancestry. And second, all human fossils that had been found in the preceding three decades had come to light through accidental discovery. No one had actually set out to find human fossils as part of a deliberate research program. This is hard for us to conceive of these days, when we are constantly reading newspaper reports of well-organized, well-funded expeditions setting out for distant lands in search of a new chapter in the story of human prehistory. The wonder of Dubois's unprecedented bravado, therefore, is not that he succeeded, but that he tried at all. In what is one of the most storied episodes of the annals of the search for human origins, Dubois found what would later be described as "the most famous, most discussed, most maligned [human] fossil."[1]

Dubois's motive for seeking the missing link in Indonesia is one of anthropology's most discussed—and largely unresolved—mysteries. At the time he set sail for Sumatra, Dubois was a young, competent anatomist. But he had had no professional involvement with evolutionary theory or with human fossils.

Born in 1858, a year before the publication of Charles Darwin's *Origin of Species,* the young Dubois quickly developed a passion for natural history. From the bedroom window of his home in Eijsden, Dubois could see the fossil-rich landscape of Limburg chalk. "I walked there many a time to collect fossils," Dubois later recalled.[2] This boyhood experience was, however, about as intimate a contact with the world of fossils as Dubois would have until his sojourn in the Dutch East Indies two decades later.

Dubois's father, a pharmacist, encouraged his son's interest in science and sent him to a state high school, rather than the Catholic school that would have been the more natural choice, because of its superior science education. Dubois flourished in the school's science-oriented atmosphere, and later commented that high school "was probably more important to me than university."[3] Nevertheless, Dubois also thrived at college, where he elected to study medicine, much to his father's disappointment, because he had hoped his son would pursue pharmacy. Dubois was a tall, strikingly handsome young man with a confident view of himself: "I had a rather good appearance and was expected to have a brilliant career," he wrote later.[4]

The young Dubois, taken in 1833, when he was an instructor in human anatomy at the State College for Teachers of Art.

Before too long, the young Dubois abandoned plans to become a practicing medical doctor and began to revel in the natural sciences, perhaps inspired by the towering intellectual talent boasted by the University of Amsterdam at the time. In 1881, Max Fürbringer, a prominent figure in the university's medical school, appointed Dubois as his assistant, thus setting Dubois on the path toward a successful career as an anatomist. And yet it was clear that Dubois was not devoted to the idea of becoming an anatomist, as a later comment of his reveals. Soon after Fürbringer offered Dubois the position as his assistant, Thomas Place, a physiologist in the same school, did the same. "If [Place] had asked me a day earlier than Fürbringer, I would have become a physiologist, which attracted me more," Dubois wrote. Then, in one of those insights into how small events can change lives and history, Dubois added the following: "Then I would never have gone to the [Dutch East] Indies."[5] And Dubois's missing link would have remained missing.

Different Views About Missing Links

The phrase "missing link" has attained mythic status in the context of human prehistory, invoking as it does some yet-to-be-discovered ancestral form lurking deep in our past. But, unfortunately, it is often misused. For instance, there is no missing link between humans and chimpanzees (our closest evolutionary relatives), despite what the popular media often state. True, humans and chimps share a common ancestor from which we both descended, but there is no missing link—in the sense of an evolutionary intermediate—between us. And there is no doubt that the author of the phrase, the German zoologist Ernst Haeckel, meant it to connote an intermediate stage of development.

Haeckel, who was born into a wealthy family in Potsdam in

1834, was a tall, handsome, charismatic man, an instinctive naturalist with a talent for communicating current ideas in science to a wide reading public. His books on zoology and evolution were translated into many languages and sold hundreds of thousands of copies. When he read the German translation of Darwin's *Origin of Species,* he instantly saw the cogency of the argument, and assumed the role of proselytizer of Darwinian evolutionary theory in Germany. In effect, he tried to do in his own country what Thomas Henry Huxley—Darwin's bulldog, as he was known—did for Darwin in England. This was no easy task because there were powerful forces aligned against Darwinism in Germany, most notably Rudolf Virchow, the founder of modern pathology and an acknowledged intellectual giant, though small and unremarkable in physical appearance. Virchow, a stickler for punctilious facts, regarded Darwin's theory as woefully speculative and misguided. He was dismayed, and disgusted, by Haeckel's enthusiasm for Darwinism. As a result of the chasm of disagreement between the two men, what had once been a student-teacher relationship filled with reverence by Haeckel for Virchow, and later one of more mutual respect when Haeckel became Virchow's assistant for a short time, declined into bitter enmity and public denigrations.

Haeckel not only was a promoter of Darwin's ideas but was also more bold, extrapolating them into the most sensitive of arenas: human evolution. Where Darwin had felt it wise to constrain any extrapolation of evolutionary theory to human prehistory to a single sentence in the *Origin,* Haeckel grasped the nettle with enthusiasm. In his *History of Creation,* published in 1868, he described evolutionary history as a series of twenty-two stages, beginning with the simplest of organisms and ending with humans. Key to the perfection of humans was the acquisition of upright walking, a large brain, and the power of speech. Based on no fossil evidence, Haeckel hypothesized that the twenty-first stage—between apes, at stage twenty, and humans, at twenty-two—would possess the full array of cherished human endowments with the exception of speech. He therefore named this creature—the missing link, as he called it—*Pithecanthropus alalus,* or ape man without speech. "The *certain proof* that such Primeval Men without the power of speech, or Ape-like Men, must have preceded men possessing speech, is the result arrived at by an enquiring mind from comparative philology (from 'comparative anatomy' of language), and especially from the history of development of language in every child as well as in every nation," wrote Haeckel.[6]

Haeckel went further, and suggested that humankind had originated on Lemuria, a putative continent that was believed to have sunk beneath the waves of the Indian Ocean, having extended from the coast of Africa to the Philippines. A pull-out frontispiece to his book showed the proposed dispersal of the twelve races of the world from Lemuria, which he labeled "Paradise." Haeckel was nothing if not imaginative.

It is hard for us today to imagine the intellectual landscape that Haeckel and others were navigating not much more than a century ago. At the time, the notion that humankind had a prehistory at all had only recently been established. And while the notion of evolution was becoming widely, if not universally, accepted, arguments raged over how it took place: it would be many decades before Darwinian natural selection would be seen as core to evolution. Even among those who accepted evolution as a fact, some prominent figures excluded humans from the picture, regarding *Homo sapiens* as fundamentally special and beyond mere materialistic processes. For instance, Alfred Russel Wallace, who devised the theory of evolution by natural selection independently of Darwin, believed spiritual intervention must have been necessary to fashion so perfect a creature as humans. No human fossils were known (or at least recognized as such) at the time. And biologists were deeply divided over the origin of human races. On one hand—the monogenist view—all human races were said to derive from a common origin. On the other—the polygenist view—races were seen as the separate products of separate origins.

The monogenist/polygenist divide has a long history, having been played out in pre- and post-Darwinian times. In the eighteenth century, for instance, the monogenist position was in the ascendant, because it accorded with the biblical account of creation. With the rise of a more scientific culture, the polygenist hypothesis gained ground and eventually dominated, because of the erroneously perceived large differences among races. When evolutionary theory became established, the divide continued for some decades, with people such as Darwin and Haeckel hewing to the monogenist view while others, such as Virchow, promulgated polygenism. In its extreme form, polygenism viewed separate races as separate species, some of which were held to be far superior to others. With this theory emanating from a Caucasian elite, there are no prizes for guessing which race was considered to be at the top of the heap.

The monogenist/polygenist debate may seem quaint or naive to us now, but that is the nature of the scientific process: at the time, the protagonists had rational reasons for espousing their separate views. In this case, however, some rather unfortunate, real-world consequences followed, as polygenism became an intellectual basis for the racism of the Nazi philosophy. More pertinent to this book is that it also influenced the way human fossils were assessed when they were eventually found, typically encouraging the opinion that the fossils had nothing to do with human ancestry and instead were mere racial variants of modern humans. This practice began with Neanderthal Man, the first specimen of which was unearthed in 1856. Even when Dubois proffered his newly discovered missing link to the gaze of the scientific world some four decades later, reverberations of polygenic thinking made for a less than enthusiastic reception of *Pithecanthropus erectus*.

Poor, Rejected Neanderthal

The fossil bones that gave Neanderthal Man his name were recovered from a cave being quarried for lime high up in a deep, narrow ravine known as the Neander Valley through which the Düssel River flows, a short distance from where it meets the Rhine at Düsseldorf, Germany. No one will ever know, but probably an entire skeleton had been entombed in the limestone sediments of that cave. All that survived the quarrymen's zeal, however, was the top of a cranium, some leg and arm bones, and other damaged parts. Overall, the fossils had a distinctive look to them: the skull bones were exceptionally thick, the eyebrow ridge unusually prominent, the limbs extremely robust, and the leg bones bowed. At the very least, Neanderthal people were stockily built, powerful individuals.

The bones—known as the Feldhofer fossil—came into the hands of Hermann Schaffhausen, an anatomy professor, who considered that here was an individual who had belonged to an apparently barbarous population from the early pages of human history, perhaps one of the most ancient races. He first presented his ideas at a meeting of the Lower Rhine Medical and Natural History Society in Bonn on February 4, 1857, which was less than three years before Darwin published his *Origin of Species*.

The immediate reaction to the Neanderthals was mixed, not least because it was impossible to say exactly what part of the geological past the bones came from, since they came from a limestone cave that was destroyed in the process of quarrying. Therefore, no one knew, or could ever know, how old the

Feldhofer fossil was. It is probably appropriate—if not encouraging—that the first human fossil to be discovered should be bedeviled with doubts about its age: *plus ça change, plus c'est la même chose!* Mostly, however, anthropologists were responding to the "uncouth" and "brutish" appearance of the Neanderthals. One German anatomist, rejecting the suggestion that the bones were ancient, said that the individual was a Mongolian Cossack of the Russian cavalry that had pursued Napoleon back across the Rhine in 1814. The Cossack, according to this expert, had become separated from his fellows, perhaps through injury, and had crawled into the cave to die. His bowed legs were clearly the result of a life on horseback. Another scholar interpreted the Neanderthal's bowed legs as the result of rickets, a condition caused by a deficiency of vitamin D that produces bone deformities. The pain from the disease had made the individual habitually furrow his brow, it was said, thus causing the prominent ridges. Still other scholars suggested that the Neanderthals might be forerunners of modern humans, or at least related to the supposedly "inferior" races, such as the Australian aborigines.

Haeckel completely ignored the Feldhofer Neanderthal in his *History of Creation*; it was only in the early years of the twentieth century that he incorporated the fossil in a human family tree. Even Huxley, an enthusiastic supporter of evolution who examined casts of the fossil bones, concluded that they represented merely an extreme form of historical human being. "These remarkable human remains belonged to a period antecedent to the time of the Celts and Germans, and were in all probability derived from one of the wild races of Northwestern Europe, spoken of by Latin writers," he wrote in his influential book *Evidence as to Man's Place in Nature.*[7] He was eager to find evolutionary links between ancient apes and modern humans, but Neanderthal Man, primitive though he was in many ways, did not fit the bill. Here Huxley was facing paleontology's perennial puzzle: what degree of anatomical difference constitutes real biological difference? Huxley was familiar with modern human anatomy, could see that Neanderthal anatomy was different in many ways, but concluded that it represented merely a primitive version of modern anatomy, not a distinct earlier form. The power of polygenism, with its emphasis on great anatomical differences between races of modern people, was subtly at work here.

Any speculation that Neanderthals might have an ancestral, rather than contemporary racial, role in human history was put to an end in the 1870s when Haeckel's adversary Rudolph Virchow pronounced the bones to be modern and pathological, not ancient. The renowned pathologist agreed with the suggestion that the Feldhofer individual had suffered from rickets. But he had been old when he died, Virchow said, and since primitive foraging societies could not support people to such an advanced age, he must have lived in a sedentary—and therefore recent—society. The bowed shape of the Neanderthal thighbones is indeed reminiscent of what occurs in rickets. But rachitic bones are thin and porous, as a result of the loss of calcium, not thick and robust as seen in the Feldhofer Neanderthal. Virchow's misinterpretation of the anatomical evidence almost certainly was the result of his own disinclination to accept any kind of evolutionary scenario and the general inclination to dwell on great anatomical variation among human races.

The biggest irony of all is that even Dubois, with his growing passion for seeking fossil evidence of human ancestry, dismissed the Feldhofer individual as having nothing to do with human ancestry. It was, he said, "not even markedly inferior to the lower races of today."[8] There's a crude saying to the effect that you should be careful you don't spit in the water because you might have to drink it! Dubois did spit in the water, and he did have to drink it, because his claims for the ancestral importance of his missing link would be greeted with dismissals similar to his own of Neanderthal.

An Obsession Takes Root

Dubois became fascinated with the subject of human origins early on in his life, and attended lectures on the topic whenever possible. His enthusiasm for discussions of human origins and for evolution in general is betrayed in a letter that the Dutch biologist Hugo de Vries wrote to Dubois shortly before he set sail for the Dutch East Indies. "I, too, always remember with pleasure the hours I spent with you, and especially [our] talks about the theory of descent and related subjects," de Vries wrote on October 28, 1887.[9] But in all the talk of evolution, whether of humans or other animals, Dubois would have heard little or no mention of fossils, for the simple reason that fossils, while interesting, were at that time considered unimportant in reconstructing evolutionary history.

The chief author of the era's fossil-free philosophical approach was none other than Haeckel, who, after all, had reconstructed the entire history of humans without benefit of a single fossil. The source of his inspiration was embryology, and in particular his famous "biogenetic law," which in very simplistic form says that in developing from a single egg to an adult, an embryo passes through all the forms of its ancestry. "Ontogeny repeats phylogeny" is the short formulation. Know the course of embryology (ontogeny) and you know a species' evolutionary history (phylogeny). Haeckel once wrote that anyone with a thorough knowledge of anatomy, embryology, and fossils, and who was open-minded enough to compare their relative importance, would have "no need of these fossil documents in order to accept the 'descent of man from the ape' as an historical fact."[10] Dubois's anatomy teacher at the University of Amsterdam, Max Fürbringer, had been a student of Haeckel's, was an enthusiastic exponent of his teacher's philosophy, and would not have encouraged Dubois to think of fossils as being important, and yet this is precisely what Dubois came to espouse. The deep reason for this remains a mystery.

True, Dubois was an outdoor type. He had loved collecting fossils as a young boy, and when he was in his late twenties he took part in excavations of human and animal bones, and flint tools, near the town of Rickholt, but they turned out—disappointingly, for Dubois—to be of recent origin. In being a fossil hunter Dubois was doing more than indulging in a healthy outdoor hobby, however. He became convinced that fossils provided the *one true way* to uncover evolutionary history. "Embryology and comparative anatomy . . . can only provide indirect proof for the existence of the close tie between Man and the animal world, only paleontology provides direct proof," he wrote in some autobiographical notes.[11] By taking this stance, the young Dutch anatomist was doing nothing less than rejecting the powerful prevailing philosophy in human origins research and replacing it with a new paradigm.

Not everyone rejected fossils as secondary in the story of human origins, however. One voice that could be taken as encouragement for seeking fossils was that of Virchow, as seen in a passage often quoted by Dubois. "Large areas of the earth are still completely unknown as concerns their fossil treasures," Virchow wrote. "Amongst these are precisely the habitats of the anthropoid apes: tropical Africa, Borneo and the neighboring islands [including the Dutch East Indies] are still totally unexplored." The final sentence of this short paragraph must have been a beacon to Dubois's obsession: "A single [fossil] discovery may change the whole state of affairs."[12] A single fossil discovery. That was Dubois's goal, and in October 1887 he set out in search of it, leaving behind a promising career with no possibility of returning to it. Why Dubois was prepared to take such a dramatic step has long been a question among anthropologists. "The riddle remains unsolved," his biographer, Bert Theunissen, wrote recently.[13]

True, Dubois had some reasons for wishing to leave Amsterdam. For instance, he had come to dislike his work in the anatomy laboratory. He detested his teaching obligations. And he had developed a deep distrust of his mentor, Fürbringer, who he believed was trying to take credit for his discoveries concerning the anatomy of the larynx in vertebrates. Dubois appears to have been wrong about Fürbringer's motives and actions, and yet the intensity with which Dubois protected what he saw as his intellectual property rights, and the depth of suspicion in which he held Fürbringer, were but the first of many such incidents in his professional life. To put it charitably, one might say that Dubois had a talent for paranoia. And although he had his reasons for leaving Amsterdam for more congenial professional surroundings, there is no obvious external reason for his burning his bridges in the process. "We shall continue to be in the dark as to the driving force behind his decision," writes Theunissen.[14]

Having made the decision, however, he had good and clear reasons for choosing the Dutch East Indies as his destination, some practical, some theoretical. The first practical reason, which all scientists can understand, was a matter of funding: Dubois needed someone to support his wildly unpredictable journey into the past. The Dutch government declined Dubois's request for such support, for the good reason that not only was the venture extremely unorthodox, but it also had vanishingly small probability of success. As a pragmatic last resort, therefore, Dubois signed up as a medical officer, second class, with the Dutch East Indies Army, with the hope of finding time for fossil hunting in between his regular duties.

In a paper he wrote in 1888, Dubois outlined his theoretical reasons—five in all—for believing that the Dutch East Indies would be a profitable place for his search. The first was that, like Darwin, Dubois believed that humans must have evolved in the tropics, where the loss of thick body hair would be tolerated. Second, Dubois noted that present-day animals typically occupy the same geographical regions as their ancestors. And since humans and apes share a common ancestor, human forebears must have lived in the tropics, where modern apes are found. Darwin made the same argument about geographical regions, but because he believed that humans are most closely related to the African apes—the chimpanzee and the gorilla—he thought that Africa must have been the cradle of humankind. Others, including Charles Lyell and Alfred Russel Wallace, believed that Southeast Asia was a more likely cradle, because this was the homeland of the orangutan and gibbon, which were said to be more closely related to humans. Dubois aligned himself with this argument.

For his third argument, Dubois pointed out that, a decade previously, the fossilized upper jaw of a chimpanzee-like ape had been discovered in the Siwalik Hills of India in 1878. This specimen eventually came to be called by the genus name *Anthropithecus.* From what was known of the fossil fauna of the Dutch East Indies, observed Dubois, it looked very much like that of the Siwalik Hills. Therefore, fossil apes—and probably fossil humans—were likely to be found in Borneo, Sumatra, and Java, too. Dubois's fourth argument rested on his assertion of similarities—based on the observations of two French paleontologists rather than his own work—between humans and gibbons and between certain fossil apes and gibbons. Tenuous at best, the putative similarities were sufficient to give further support to Dubois's Indonesian venture. The fifth reason simply reflected contemporary experience in the search for human fossils: such fossils had always been found in caves. There are lots of caves in the Dutch East Indies, especially in Sumatra.

So, Sumatra it was to be.

5

Dubois's Story: Link No Longer Missing

DUBOIS was a groundbreaker in human origins research, primarily because he was the first person to set out on a deliberate search for fossils of human ancestors. Recall Rudolf Virchow's line: "A single discovery may change the whole state of affairs." We can imagine that this tantalizing sentiment must have been ringing in Dubois's ears as he neared the shores of Sumatra on that eleventh day of December in 1887. Having convinced himself, if not yet others, that this Dutch East Indies island would contain a trove of ancient fossils, including the missing link between apes and humans, Dubois was anxious to begin the search. But he had to stay his enthusiasm for half a year while the medical duties for which he was being paid occupied virtually all his time. In May 1888 Dubois managed to get himself transferred to Pajakombo, in central Sumatra. There he would have not only readier access to the caves he dreamed about but also lighter medical duties, thus giving him time to explore the caves.

Dubois began his investigations that July, and within a month he had unearthed the bones of orangutans, gibbons, rhinoceroses, deer, elephants, and various other beasts from the first cave he explored, named Lida Adjer. It seemed a brilliant beginning, as it offered proof that fossils were indeed to be found on the island. By this time Dubois's lengthy article on the paleontological promise of the Dutch East Indies, and Sumatra in particular, was in press in the journal *Natuurkundig Tijdschrift.* In addition to the five scientific arguments in the article (outlined at the end of chapter 4), Dubois appealed to national prestige with the following passage: "Will the Netherlands, which has done so much for the natural sciences of the East Indies colonies, remain indifferent where such important questions are concerned, while the road to their solution has been shown?"[1]

The confection of science and national pride, along with the demonstration that he could find the fossils he talked about, prevailed. In March 1889 the Dutch government relieved Dubois of his medical duties and transferred him to the Department of Education, Religion, and Industry, specifically to pursue his fossil-hunting plans. He was assigned two members of the engineering corps, Franke and Van den Nesse, and fifty forced laborers (convicts, in reality) to carry out the excavation work. Dubois, ecstatic with this vote of confidence, looked forward with certainty to building on his initial success.

What a huge disappointment awaited him. The expedition proved to be a disaster. Within a year Dubois was in despair. "Everything here has gone against me," he wrote in a letter to his friend F. A. Jentink, director of the National Museum of Natural History in Leiden. His workers were "as indolent as frogs in winter." He discovered that the local people were being secretive about the whereabouts of caves, believing that the government might try to extract the gold and other minerals that they wanted for themselves. He hated spending "weeks on end" in the forest with little food and shelter.[2] He suffered repeated bouts of fever. He had to dismiss one of the two engineers for incompetence; the other one died. Seven of the laborers ran away or were dismissed, and half of the rest fell sick. It was virtually impossible to get around from cave to cave because the mountainous terrain was so precipitous and thickly forested. Worst of all, the few fossils he managed to recover proved to be of a very young age, which meant there was no chance of finding human fossils that had anything to do with human antiquity. There was no missing link to be found in Sumatra. Dubois, in his quarterly reports on the expedition, was careful to avoid describing the expedition as the fiasco it so obviously was. Instead, he said that his work on Sumatra was complete, and that a switch to Java would be the wisest next step.

Dubois was not just retreating from Sumatra while declaring victory; by this time there were already indications that Java might indeed hold what he searched for. First, recent geological work had shown deposits on the island to be as ancient as Dubois hoped and required. Second, and more dramatic, was the discovery in October 1888 of a fossilized human skull in a rock shelter near the hamlet of Wadjak, in southeastern Java. While it was not the ancient missing link Dubois sought, neither was it fully modern. Dubois considered it to be a primitive race of modern humans, as he had with the Neanderthal, and a portent of more important finds to come.

On April 14, 1890, the Dutch government gave Dubois permission to transfer his work to Java. The young Dutchman spent a month in fevered preparation for the transfer, full of anticipation of what lay in store for him on the neighboring island to the east.

JAVA, "THE CRADLE OF HUMANKIND"

Dubois began his work in Java in traditional mode, by concentrating his searches in caves. As on Sumatra, two officers from the engineering corps were assigned to direct excavations, again carried out by a team of convicts. In contrast with the situation on Sumatra, however, the officers, G. Kriele and A. De Winter, were competent and the laborers effective. Dubois established a base at Toeloeng Agoeng, in southeastern Java, from where he visited the excavations occasionally, but more often he received the fruits of their work at his house, where he cleaned, identified, and documented them after releasing them from their teak-leaf wrapping. The cave sites yielded only small rewards, however, and so Dubois soon directed the exploration to open sites, too. These included deposits in the Kendeng Hills and along the banks of the Solo River. The timetable of work was determined by the weather, because in the rainy season (December to May) the swollen river flooded excavation sites there, and in the dry season (June to November) a thick carpet of teak leaves covered the ground in the hills.

The switch to noncave sites proved to be inspired, and before long a veritable mountain of animal fossils was building, including specimens of a primitive elephant (Stegodon), hippopotami, rhinoceroses, hyenas, big cats, and many others. Eventually these fossils numbered more than 12,000 and occupied more than 400 cases when they were shipped back to Holland. More important than their number and diversity, however, was their likely age. From the kinds of animals in the collection and their stage of evolution, Dubois could see that they were very similar to those in certain sedimentary levels in the Siwalik Hills of India. This meant that they hailed back to the end-Pliocene/beginning-Pleistocene era, which, in today's terms, means that they lived and died some 1.5 to 2 million years ago. This was exactly what Dubois wanted, because he believed that the human line had begun to evolve by this time. The only question remaining for Dubois was, When would human remains begin to come to light in his excavations?

Dubois didn't have to wait long, and his response to this historic find illustrates the awkward mental framework in which anthropologists of the time found themselves. On November 24, just a few months after Dubois began his Javan venture, a fragment of lower jaw was discovered among animal fossils of great age at a site called Kedung Brubus, in the hills south of the Solo River. Dubois recognized the fossil (part of a chin, bearing gaping sockets where the canine and first and second molars once lodged) as being humanlike yet primitive. He suggested that the jaw had belonged to a member of our own genus, *Homo,* but shrank from declaring it *Homo sapiens,* although he did talk about the individual as being from a primitive race of humans. In a letter to his friend Jentink, he wrote that the jaw was "not a recent [type] . . . although nowhere near being an anthropoid type."[3] Here we see once again the nineteenth-century view that in order for a specimen to be regarded as ancestral to humans, it had to be extremely apelike. Anything ancient that had clear human affinities was included in the circle of *Homo sapiens,* with the rider that it was a primitive race. The result was that in the late nineteenth century, no early evolutionary form of modern human had been recognized, even though some existed in the form of Neanderthals.

The following August, Dubois's crew began developing the excavation on the Solo River, near the village of Trinil, that we described in chapter 2. Back in Dubois's time, the region was more thickly forested and less farmed. The excavation on the curve of the river was circular, about 14 yards in diameter. Within this confined space, as many as fifty people sometimes labored, cutting back through time as they cut down through sediment layer after sediment layer. One can only imagine the sight and sound of the activity when the crew was in full swing in this otherwise tranquil spot! At about 17 yards below the modern surface, the excavation reached a hard layer, about a yard thick, that contained volcanic rock and many fragmented fossils. Historic discoveries were about to be made.

The first, in September, was relatively modest, a cheek tooth (the third molar) from a large primate. Dubois, considering the tooth a relative of the fossil chimpanzee that had been found in the Siwalik Hills in 1878, gave it the same genus name, *Anthropithecus,* or manlike ape. Despite its being a single tooth, the find encouraged Dubois that he was on the right track. A month later this was confirmed, with the unearthing of one of the most famous of all human fossils, the Trinil skull cap. Although it lacked a face and the lower part of the cranium, the skull cap was still quite revealing. It was long and low, not domed as in modern humans, and had pronounced bony ridges above the eyes, which is somewhat apelike. Nevertheless, the cranium had obviously housed a brain considerably bigger than any ape boasts. Despite calling his find a chimpanzee, Dubois recognized that he possessed a very important specimen, one that was "truly a new and closer link in the largely buried chain connecting us to the 'lower' animals," as he wrote to a friend.[4] With so spectacular a find following so closely on the discovery of the molar tooth, Dubois must have expected, or at least hoped for, more to follow swiftly. If so, he was to be disappointed, because the season ended with no further humanlike finds.

The next excavation season began in May 1892, with Dubois's crew opening up another excavation upstream from the first one. Three months later, a second spectacular find was made, one that shifted Dubois's view of the nature of the skull cap, and which, together with the skull cap, completed the discovery of, in von Koenigswald's words, "the most famous, most discussed, most maligned fossil."[5] The new find was a complete and pristine left thighbone, or femur, marred only by an area of secondary bone growth near the top of the thigh, which was most likely the result of injury and subsequent recovery. (Dubois later speculated that the injury was probably from an arrow or spear wound, but given our current knowledge of the age of the Trinil fossil and the age of the oldest known projectiles, this now seems unlikely.) The fossil thighbone as we see it today lacks a few slivers of bone from its shaft, an absence that is explained by a letter from Kriele to Dubois, on September 7, 1892: "De Winter told me that those small pieces missing from [the thighbone] were blown away by a strong wind while being glued together on a [teak] leaf, and could not be found again."[6]

Marginally incomplete though it therefore was, the character of the thighbone was unmistakable: "It shows a striking similarity to this supporting bone of the human body,"[7] Dubois wrote in his third quarterly report for 1892. The bone is indeed very similar in anatomy to a modern human counterpart, with the exception that the fossil thighbone is thicker. "One can say with certainty that *Anthropithecus* of Java stood upright and moved like a human."[8] As a result of this insight, Dubois gave his missing link a species name for the first time, *erectus,* the full name being *Anthropithecus erectus,* or upright man-ape. This discovery, Dubois concluded, demonstrated that, "as some people have suspected, the East Indies was the cradle of humankind."[9]

ENTER *PITHECANTHROPUS*

There are many myths in the popular story of Eugène Dubois and the man-ape of Java. One of them is that, having discovered the missing link that Ernst Haeckel predicted must exist, Dubois named the creature *Pithecanthropus erectus,* using the genus name Haeckel had coined for his hypothetical human ancestor, *Pithecanthropus alalus,* or speechless ape-man. Although Dubois would soon give the fossils the genus name *Pithecanthropus,* ape-man, his initial choice was *Anthropithecus,* man-ape. As Dubois's recent biographer, Bert Theunissen, has convincingly argued, Dubois made the switch to *Pithecanthropus* for reasons other than recognition of Haeckel's prescience.

The Trinil skull cap was filled with hard rock matrix when it was unearthed, as is typical with fossil crania, the Mojokerto child included. This makes calculation of the creature's brain size difficult, though not impossible. In his initial calculation of the original Java Man's brain capacity, Dubois made an error, concluding that it was about 700 cubic centimeters. Now, although this is 1.75 times the size of a chimp's brain, it is only half that of a modern human brain. The humanlike anatomy of the thighbone convinced Dubois that Java Man walked upright, as humans do, but the calculated brain size indicated that he was more ape than human. The name *Anthropithecus,* man-ape, closer to apes than humans, seemed appropriate.

At some time between the end of November and the beginning of December of 1892, Dubois discovered his error. When he did the recalculation, the brain size came out substantially larger, some 900 cubic centimeters, which made Java Man much more humanlike than Dubois had initially believed.

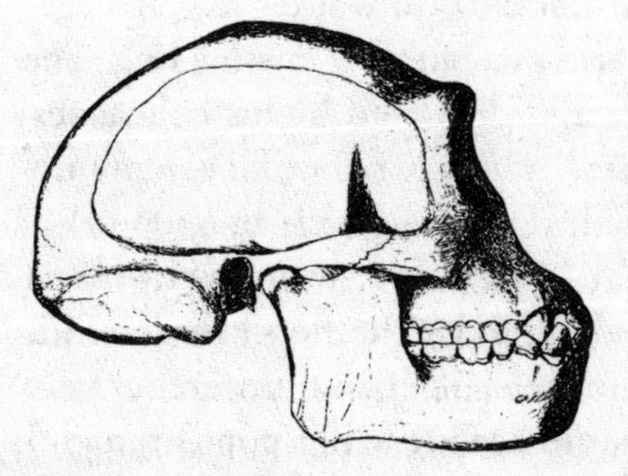

Dubois's first reconstruction of Pithecanthropus, in 1896. The image is a mixture of human features (high, vaulted skull) and ape features (protruding jaw), reflecting Dubois's belief at the time that Pithecanthropus was neither human nor ape, but something in-between.

Clearly, the name *Anthropithecus* was inappropriate, because Java Man was now seen to be more human than ape. With a name change necessary, *Pithecanthropus,* ape-man, must have seemed apt to Dubois, although he left no contemporary reasoning for the choice. But it surely did not escape Dubois's attention that choosing *Pithecanthropus* might help recruit Haeckel's support in the next step of the venture, which was to convince the anthropological world that he had indeed found the missing link between humans and apes, just as he said he would. Dubois did, however, write a letter to Haeckel a few years later: "I should like to tell you how happy I am to be able [by the discovery of *Pithecanthropus*] to express my gratitude for the influence which you . . . have exerted on the whole course of my life."[10]

The Trinil thighbone was the last significant humanlike fossil that Dubois was to find at Trinil, or anywhere else for that matter. For reasons of health, Dubois and his wife left Java for the Netherlands in 1895, never to return. With the discovery of two pieces of fossilized bone—a skull cap and a thighbone—Dubois was poised to revolutionize the anthropological world, but it would be a long, rough ride.

The Battle Begins

The second popular and persistent myth about Dubois is his supposed reaction to the anthropological world's largely negative assessment of *Pithecanthropus* as a putative missing link. The story goes as follows. Angry and dismayed by his colleagues' recalcitrance, Dubois withdrew from the scientific community and for more than two decades hid the fossils to keep other scholars from studying them. (There are several versions of where he stored them, including under the floorboards of his dining room and in a box in the attic. These differences are perhaps an indication that we are dealing with a popular myth.)

the Mojokerto child have enjoyed, had she lived? In answering these questions, we gain a deeper appreciation for what kind of animal we—as members of *Homo sapiens*—are. We gain a clearer perspective of our evolutionary heritage and of our relationship to the rest of nature.

In this chapter and the next two, we will explore these questions by looking, first, at the early stages of human prehistory and at the way anthropologists have thought about them; second, by sketching out what we know about the life and times of *Homo erectus*; and third, by exploring what is known about brain expansion and its impact on childhood and on our ancestors' technological and language abilities. Our aim is to sketch a clear image of the evolutionary context of the main issue of our book: namely, when did our ancestors first extend their range beyond Africa, and what kind of creature was it that made this journey, on its way to becoming people like us?

Evolutionary Stories, and the Roots of Humanity

Humans have been asking questions about their origins and their place in nature ever since true self-awareness began to flicker in the human mind, we can be sure of that. Every society in the modern and recent world has, or had, an explanation of how its people came to be, encapsulated in creation myths, stories that explain its people's place in the world. Recorded history stretches back only half a dozen millennia, when writing was invented. Creation myths abound here, captured like birds on the wing, in forms of writing or hieroglyphics. In the time before writing—literally, prehistory—creation myths existed, not encrypted on parchment or clay, but in people's minds, to be shared as stories told and heard. And, probably, as images painted and engraved on cave walls or rock shelters. It's all but large brain, a large human-shaped (as opposed to ape-shaped) body, and long lower limbs and relatively short upper limbs. These physical differences contributed to the species' ability to be the first human species to expand its range beyond Africa, where the human family had been born, and to establish homelands in Asia and Europe. It was the first to obtain a significant amount of meat in its diet through hunting, not just through scavenging. No previous human species had routinely established what might be called home bases, temporary campsites in the endless round of a nomadic lifestyle. No previous species could claim to manufacture complex stone tools into complex, preconceived shapes such as hand axes. *Homo erectus* was the first human species to use fire. It may have been the first time in human history that human babies had an extended childhood, similar to ours, a period of care and attention after weaning and before maturity when they could not survive without their parents. *Homo erectus* individuals may also have for the first time experienced an adolescent growth spurt, signaling an extended adolescence that was critical for learning complex cultural behaviors.

There is no question that the Mojokerto child was born into a way of life that was very special in human prehistory, even though her untimely death—at the tender age of about five—robbed her of an opportunity to experience it. Nothing like this way of life had existed previously; it was the foundation upon which the lifeways of fully modern humans—people like us—were later built.

The time of the Mojokerto child and her species was therefore a pivotal point in the evolutionary transformation of an apelike ancestor into modern humans, having accrued many physical and behavioral features that we would recognize as being humanlike, although not fully human. We want to ask, What kind of animal was *Homo erectus*? What kind of life would

Then, it is said, in frustration, anger, and a touch of madness, Dubois declared the fossils to be nothing more than those of an extinct giant gibbon, having nothing to do with human ancestry.

It is true, as we shall see, that Dubois's proposition found little support among professional anthropologists, some of whom dismissed *Pithecanthropus* as being merely a primitive human race, not a human ancestor, just as Dubois had done with Neanderthal. But the rest of the story, about Dubois's supposed withdrawal and craziness, is apocryphal, as Bert Theunissen has so ably argued.

Dubois's first major task after the 1892 excavation season was to prepare a description and interpretation of *Pithecanthropus.* There were two main points. First, he asserted that it would be foolish to doubt that the three fossils (tooth, skull cap, and thigh) belonged to the same individual, because, he reasoned, they were found in the same sedimentary layer and within a few yards of one another. Second, he claimed that *Pithecanthropus* was neither ape nor human but rather an evolutionary link between the two. "*Pithecanthropus erectus* is the transitional form which, according to the theory of evolution, must have existed between Man and the anthropoids; he is Man's ancestor," Dubois wrote in his dissertation.[11] Dubois was hampered in writing the dissertation, which was published in 1894, because he had few fossils or fossil casts with which he could compare his *Pithecanthropus,* and his access to scientific literature was limited. These handicaps might have contributed to the dissertation's not being as thorough as it might otherwise have been, but they surely do not account for the extent and tenor of the criticism that was soon to be heaped on his central two points.

The tide of criticism began to flow in 1893, following the publication of a translation of Dubois's third quarterly report for

1892, which described the find as *Anthropithecus erectus.* It swelled with the publication of Dubois's thesis in late 1894, in which the name *Pithecanthropus erectus* made its first public appearance. And it surged to floodlike proportions when Dubois returned to Europe and embarked on a proselytizing odyssey through, first, Leiden, then Liège, Brussels, Paris, London, Cambridge, Edinburgh, Berlin, and Jena. Dubois was soon seething at being congratulated on making an important discovery while having every aspect of his interpretation attacked. Rudolf Virchow taunted Dubois with a paraphrase of his own words, saying, in effect, that it would be foolish *not* to doubt that the three fossils belonged to the same individual. Dubois was to hear this criticism wherever he went, and in truth, it is valid. Dubois had failed to be punctilious about the exact location of the discovery of each fossil, primarily because he was usually absent from the excavation. And, scattered as they were over many tens of square yards, the fossils were part of a bone accumulation that was deposited by flowing water. It is impossible to be certain, as Dubois was, that the fossils once were part of one individual.

This was bad enough for Dubois. Worse, however, were the criticisms of his interpretation of the bones themselves, which included every possible counterview. The skull is that of an ape, some said. No, it is fully human, others retorted. The thighbone is obviously from a modern human, some averred. Obviously an ape's, opined others. Few, with the exception of Haeckel, the American paleontologist Othniel C. Marsh, and, later, the French anthropologist Léonce Manouvrier, accepted Dubois's core point, that *Pithecanthropus* was a transitional form and part of human ancestry. It is interesting to note that the direction of criticism, and the preferred alternative interpretation, often ran along national lines. For instance, British anthropologists championed the notion that the skull was human, while German scholars lined up behind the ape skull position.

Fifty years later, Ralph von Koenigswald, who followed in Dubois's Javan footsteps, described this cacophony in his popular book *Meeting Prehistoric Man.* "A primitive man or a specialized ape? No other paleontological discovery has created such a sensation and led to such a variety of conflicting scientific opinions."[12] Think how frustrating it must have been for Dubois to face such a variety of conflicting scientific opinions. And think how difficult it must have been to respond to them. The diversity of views reflected the state of anthropology at the time, which was governed more by preconceived ideas than by a solid foundation of accepted facts, for the good reason that there were very few of them. (To some degree, this is still true today.)

By the end of the nineteenth century, close to eighty articles and books had been published on the Trinil fossils and their place in human history. Dubois remained adamant throughout that *Pithecanthropus* was a transitional form, a member of our ancestral lineage. While counter-arguments persisted, anthropological opinion slowly began to swing in Dubois's favor, often after individual scholars had had an opportunity to see the fossils at first hand. More important, however, was that human origins research itself was beginning to transform as a result of the *Pithecanthropus* debate. The strictly anthropological approach of the nineteenth century, with its later emphasis on great differences between the races, began to give way to a more evolutionary approach. In this newly emerging perspective, anatomical differences were more readily accepted as implying evolutionary differences, not simply racial differences. And for the first time, fossils were coming to be regarded as important evidence in the attempt to understand human prehistory. As the American paleontologist Edwin Drinker Cope said at the time, "The ancestry of man is a question to be solved by paleontology."[13] That was new.

One example of the changing perspective came in May 1897, when Dubois invited the Strasbourg anatomist Gustav Schwalbe to come to Leiden to study the Trinil fossils. When Schwalbe saw the fossils at first hand—he had previously seen only casts—he became convinced that it was time to take another look at the long neglected Neanderthal fossils. He soon came to the conclusion that there was an evolutionary sequence, beginning with *Pithecanthropus,* leading through Neanderthals, and continuing on to modern humans. In effect, the Neanderthals had been rediscovered, and evolutionary models were beginning to emerge. Such was the impact of Dubois's visionary work.

Despite this groundswell of intellectual change, Dubois remained aloof from his colleagues. And at the Fourth International Congress of Zoology in Cambridge, England, in 1898 he still felt the need to defend himself against his critics, even though their number was diminishing. The Cambridge presentation was Dubois's last major contribution to the *Pithecanthropus* debate. Two years later, for the World Exhibition in Paris, Dubois supervised the construction of a life-size model of *Pithecanthropus* which emphasized the humanness of the head and the apeness of the limbs. Here, in full-fledged tangible form, was Dubois's statement that *Pithecanthropus* was transitional. Following the exhibition, Dubois withdrew himself from the debate and his fossils from public and professional view.

The folk wisdom for explaining Dubois's dramatic action is that the honorary degrees, medals, and diplomas he received were not what he craved. Nothing less than emphatic agreement that *Pithecanthropus* was a transitional form between apes and humans would be acceptable to him, having for so long been the brunt of doubt and scorn. In the absence of complete acceptance, Dubois spitefully withheld further access to the prize of Java, the story goes. Once again Bert Theunissen offers a different, and more plausible, view. He admits that Dubois "had an infinite capacity for annoyance with any scientist whose viewpoint differed from his own," and that "his suspicious nature, verging on the paranoid, sometimes caused him to suspect dark motives behind colleagues' disagreement with his views."[14] But if the conventional explanation of Dubois's withdrawal is correct, why, asks Theunissen, would he have withheld the fossils from friends as well as foes? And why would he do it with the intellectual tide turning his way?

Theunissen points to the professional circumstances to which Dubois returned in the Netherlands in 1895. First, he had accepted a not very prestigious professorship at the University of Amsterdam, which meant that his present and future status derived primarily from his possession of the anthropological world's first known missing link. Second, he had given Schwalbe access to the fossils in 1897 and had seen him build his reputation on the important interpretation he developed from them. "Dubois must have realized that, if he continued in this way, there would soon be little credit to be gained by producing a new description," suggests Theunissen. "He probably drew a lesson from this affair and decided to keep the fossils to himself until he had written about them."[15] The loss of intellectual property rights that Dubois believed had happened with his supervisor Fürbringer a decade earlier obviously still stung. He was determined not to let it happen again. Dubois was determined that it was *his* name that would be associated with *Pithecanthropus,* and no one else's.

The anthropological world had to wait many years before Dubois finally produced the new description of the fossils to which he was committed, however, because he was immersed in other, more interesting quests.

An Argument from Brains

As an anatomist, Dubois was primarily interested in brains, specifically the size of a species' brain in relation to the size of its body. These days the subject is a thriving issue in modern evolutionary and behavioral biology, but a century ago Dubois was a lonely pioneer. Dubois actually published far more on brain evolution than on *Pithecanthropus,* for which he is far better known. The two subjects were not separate in Dubois's scientific life, however, because he said an understanding of brain size was crucial to determining the intermediate status of *Pithecanthropus.* "It was to obtain a better insight into this new organism that, soon after the discovery, I undertook the search for laws which regulate cerebral quality in Mammals, a study which indeed furnished evidence as to the place of *Pithecanthropus* in the zoological system," he wrote in 1935, five years before he died.[16]

By 1928, Dubois had developed a theory (erroneous, as it turned out) about the evolution of brain size increase in mammals, in which, with each major evolutionary advance, the brain doubles in size relative to the body. He based this idea on his understanding of embryology. The brains of great apes are a quarter the human size; the brains of carnivores and hoofed herbivores are an eighth; the brains of rabbits are a sixteenth; and so on. In this scheme, Dubois recognized that there was a gap: something should fit halfway between the great apes and humans, something with a brain half the size of the human brain. For *Pithecanthropus* to be the missing link that Dubois believed it was, it would have to slide in between the two, at precisely half the relative brain size of humans. The problem was, the brain of *Pithecanthropus* measured 900 cubic centimeters, that is, two-thirds the size of a modern human brain, not half. This would put the Java fossil creature outside the evolutionary progression toward *Homo sapiens.*

Dubois therefore indulged in an interesting line of reasoning that, to him, restored the status of *Pithecanthropus* as a human ancestor. If *Pithecanthropus* had a humanlike body, then its brain would indeed be too big for that of a direct human ancestor. But if its body had been much larger, say 220 pounds rather than the human average of 132 pounds, then its brain size relative to its body size would be reduced. In fact, it would then fit the required halfway position between great apes and humans, and would provide a jumping-off point for the evolution of true humans.

To bolster this argument, Dubois developed the view that although the *Pithecanthropus* was human in shape, the proportion of the length of the femur to the length of the trunk was different in *Pithecanthropus* from the proportion in modern humans.

Dubois in 1928, twelve years before he died.

By different, he meant more apelike, with shorter legs compared with the trunk. The reason for this, he said, was that locomotion in *Pithecanthropus* was different from the way we walk, and that it "cannot have been exclusively, perhaps not even chiefly, on the ground." (Notice that Dubois had changed his opinion about locomotion in *Pithecanthropus,* which he initially said was virtually like that in modern humans.) This apelike body build would then give *Pithecanthropus* the body weight that would make the ratio of its brain weight halfway between that of modern humans and modern apes. "From all these considerations it follows that *Pithecanthropus erectus* undoubtedly is an intermediate form between Man and Apes," he wrote.[17] Presto, the role of missing link for *Pithecanthropus* is rescued! If this argument sounds circular, that is because it is.

Because Dubois applied the name "Giant Gibbon" to this creature, many people took it to mean that, in a fit of pique or madness, he no longer considered his *Pithecanthropus* to be linked to human ancestry. Not so, but the myth persists. In common with other anthropologists of the time, Dubois believed that the human stock was rooted in some kind of gibbonlike ancestor. By describing *Pithecanthropus* as a giant gibbon, Dubois simply meant that it was closer to gibbons than to humans in body form. And, he pointed out, gibbons and humans share many anatomical features that relate to humans' habitually and gibbons' occasionally upright mode of walking. If there is any doubt of Dubois's views over this, it should be relieved by a statement he wrote in 1932: "I still believe, now more firmly than ever, that the *Pithecanthropus* of Trinil is the real 'missing link.'"[18]

In the early 1930s, four separate fragments of *Pithecanthropus* thighbone were discovered in the boxes of animal fossils that Dubois had shipped from Java to Holland. This was not the first time important fossil "finds" had been made in boxes in museum basements, nor was it the last. In this case, the finds served to bolster Dubois's view of the genuine species status of *Pithecanthropus erectus.* Each of the partial thighbones displayed features that Dubois had seen in the original Trinil bone, including extra thickness compared with modern human bones. This commonality of anatomical form confirms "the right of *Pithecanthropus erectus* to exist as a real species, of a separate genus, with a particular organization," he wrote.[19] Without sinking into the swamps of taxonomic classification, we should clarify Dubois's concern with recognizing *Pithecanthropus* as "a separate genus with a peculiar organization." When he talked of his Java Man as being neither ape nor human but rather a transitional form, he insisted that this should be recognized in formal terms of classification. For instance, all human-related species (living and extinct) belong to the taxonomic family Hominidae. Similarly, all ape-related species (living and extinct) are in their own family, the Pongidae. For Dubois, *Pithecanthropus* belonged to neither one family nor the other. It was, he said, the sole occupant of a separate family, the Pithecanthropoidae. This was the only status that could justify its being recognized as a missing link, Dubois insisted.

We have seen that Dubois behaved like a jealous lover in his protection of his missing link from the first day of its discovery. This has occurred many times in human origins research, not least because important discoveries are made infrequently and bestow a certain prestige on their finders. Thankfully, it is a less common behavioral trait in the profession now than it once was. During the four decades of his possession of *Pithecanthropus,* Dubois had felt the need to defend his charge in public, usually because he believed his fossil's status as an evolutionary transitional form was misunderstood. The above defense, of "the right of *Pithecanthropus erectus* to exist as a real species," was prompted by a new challenge, however. New discoveries of fossil finds in China and Java were threatening to diffuse the importance of his Trinil *Pithecanthropus* by absorbing it into other ancient populations, which belonged to the family Hominidae. In which case, *Pithecanthropus* would be a member of Hominidae, too, and no longer his cherished missing link. Or so he thought.

Ossified Bones, Hardened Opinions

The revolution in anthropology ignited by Dubois's work had prompted scholars to follow in his footsteps and go out in search of human fossils. For instance, a German expedition, led by Margarete Leonore Selenka, pitched a camp on the opposite side of the river from Dubois's Trinil site and spent an enthusiastic season unearthing hundreds of fossil bones, but no *Pithecanthropus.* The site of the camp was easy to find, von Koenigswald wrote in 1956—it was "strewn with innumerable beer bottles, testifying to the expedition's thirst."[20] With a little effort, it is possible to find beer bottles there today. Other human fossil seekers were at work in South Africa, where the first human much older than *Pithecanthropus* was found by Raymond Dart in 1925; and in China, where, in 1927, the Canadian Davidson Black recognized humanness in two isolated teeth, from the cave site of Zhoukoudian in Dragon Bone Hill, near Peking, now Beijing. On this slenderest of evidence, Black named them a new species of human, *Sinanthropus pekinensis,* or, more popularly, Peking Man. This discovery was to be the beginning of the end for *Pithecanthropus.*

Over the next few years, several skull caps were recovered from Zhoukoudian, eventually making it one of the richest human-fossil sites ever discovered. Sadly, Black died prematurely of a heart attack while sitting at his desk, in March 1933. His place was taken by the German anatomist Franz Weidenreich.

In Black's view, there were many similarities between *Sinanthropus* and *Pithecanthropus,* with the Chinese fossils perhaps being more advanced, in the direction of Neanderthals. There was no doubt in Black's mind that *Sinanthropus* belonged in the human family, a member of the Hominidae. Weidenreich went even further, suggesting that Chinese and Javanese were so similar to each other that they should be lumped in the same group, within the Hominidae, of course.

Dubois was outraged, because such a suggestion would rob Java Man of its uniqueness. He agreed that *Sinanthropus* was part of the human fold, a form of Neanderthal, he said, adding condescendingly that it "may be an interesting new race."[21] But as for the notion of a familial link between *Sinanthropus* and *Pithecanthropus*—never! We can see now why Dubois wrote his 1935 article, "On the Gibbon-like Appearance of *Pithecanthropus erectus*": he wanted to emphasize the differences between the Chinese and Javanese fossils, to preserve the uniqueness of *Pithecanthropus.* But new pressure was soon to come from Java itself, when von Koenigswald began prospecting there for human fossils.

Von Koenigswald's first discovery was the Mojokerto child, in 1936. "That we were dealing with a human skull was clear from the outset," he wrote. He therefore named it *Pithecanthropus modjokertensis.* "A protest immediately arrived from Dubois with the statement that his *Pithecanthropus* was not human, and therefore this cranium could not be a *Pithecanthropus,*" recalled von Koenigswald. To be diplomatic he renamed the child *Homo modjokertensis,* but he still referred to it as *Pithecanthropus* in discussions. From von Koenigswald's writings, it is clear that Dubois made a pest of himself, with a steady rain of complaints, making the solving of the "riddle of *Pithecanthropus*" somewhat tedious. "Only a new *Pithecanthropus* skull could help us with this," von Koenigswald wrote.[22]

His wish was fulfilled the following year, at Sangiran, in East-Central Java. At the beginning of September, von Koenigswald heard from his field director that a small piece of human skull had been found at Sangiran. "I took the night train to Central Java the same evening, and the next morning I was at the site," wrote von Koenigswald. The Javanese had become enthusiastic fossil collectors, usually for medicinal reasons, but their help could be recruited for some financial consideration. "I had . . . promised 10 cents for every additional piece belonging to the skull," he wrote. "Consequently we were forced to buy an enormous mass of broken and worthless dental remains and throw them away in Bandung," von Koenigswald continued; "if we had left them at Sangiran they would have been offered to us for sale again and again!"[23]

Very soon, however, the Javanese began to bring useful fragments, pieces of human skull, to von Koenigswald. He was delighted because it rapidly became apparent that the fragments, though small, constituted a virtually complete skull. Delighted, that is, until he saw what was happening. "Too late I realized that my opportunist brown friends were breaking up the larger pieces behind my back, in order to get a bigger bonus." Nevertheless, it was a great find, and a celebration was in order. "We wound up the eventful day with a feast," he recalled. "We distributed rice and salt, laid on a *gamelang* orchestra, and the village's three dancing girls even put in an appearance. The latter were heavily made up and no longer in their first youth; they were so-called *ronengs,* who lived outside the village with an old woman and were celebrated for their love charms and love potions."[24]

The skull proved to be crucial: it contained the region of the ear hole, and the joint with the lower jaw. The anatomy of these regions is distinctive between humans and apes. And the anatomy of the new Sangiran skull was unequivocal: "This find . . . proved at last that *Pithecanthropus* was human." Naively, von Koenigswald sent a photograph of the reconstructed skull to Dubois, expecting that "he would share my joy that the problem had finally been solved, even though it didn't confirm his current opinion." Instead of receiving a joyful reply from Dubois, von Koenigswald was sent by a friend in Holland a copy of an article by Dubois, which included a photograph of the Sangiran skull and a statement that the skull had been incorrectly reconstructed. "He indirectly accused me of having faked the skull," fumed von Koenigswald.[25] Dubois said that von Koenigswald had made the skull too small and the vault too

Ralph von Koenigswald, right with pith helmet, holds a fragment of *Pithecanthropus* II, a relatively complete cranium that was pieced together from fragments. The photograph was taken at the site in the Sangiran region where the fossils were found, in September 1937. With von Koenigswald are some of his helpers, and young spectators. (R. von Koenigswald, courtesy of the Department of Library Sciences, American Museum of Natural History)

flat, thus making it look more like *Pithecanthropus* than he, Dubois, believed it to be.

Von Koenigswald sent a fierce letter of protest to the Royal Dutch Academy of Science. The letter was not published, but Dubois was prevailed upon to climb down a little, which he did in a way that von Koenigswald found even more insulting. "[Dubois] did not wish, he wrote, to insinuate that I had intentionally altered the shape of the cranium in putting it together," wrote von Koenigswald; "the skull consisted of so many fragments that even he [Dubois], with fifty years of experience in this field, could probably not have reconstructed it in its original form."[26] This was nonsense, observed von Koenigswald; the reconstruction was "child's play" because the cranial bone is so thick.

For von Koenigswald, the riddle of *Pithecanthropus* was solved: it was in the human camp. He and Weidenreich spent the next two years comparing in detail the anatomy of *Pithecanthropus* and *Sinanthropus,* mostly by correspondence but also through a visit by von Koenigswald to Beijing in 1939. As a result, they concluded that the two fossil groups represented the same evolutionary stage, and that they were "related to each other in the same way as two different races of present mankind."[27] Dubois would hear none of this; he launched a further series of attacks, in a trilogy of papers published in the journal *Proceedings.* The final paragraph of the third paper is poignant, and says as much about the author as about the intended audience:

> It is most regrettable, that for the interpretation of the important discoveries of human fossils in China and Java, Weidenreich, Von Koenigswald and Weinert were thus guided by preconceived opinions, and consequently did not contribute to, on the contrary impeded, the

> advance of knowledge of man's place in nature, what is commonly called human phyletic evolution. Real advance appears to depend on obtaining material data in an unbiased way, such as in the *Pithecanthropus* fossils and instructive material about the phylogenetic growth of the brain.[28]

Dubois died two weeks after these words were published, on December 16, 1940. The same firm conviction that he was right, in seeking the missing link in the Dutch East Indies, was eventually to do him disservice, in making him blind to new data and new interpretations. The British anthropologist Sir Arthur Keith wrote in an obituary notice that Dubois was "an idealist, his ideas being so firmly held that his mind tended to bend the facts rather than alter his ideas to fit them."[29] Incidentally, Keith was to bend a few facts in his time, over Piltdown Man, but that's another story.

The final transformation of Java Man came in 1944, when the Harvard evolutionary biologist Ernst Mayr, seeking to rationalize what by then had become a menagerie of genus and species names in the world of anthropology, included the Javan and Chinese fossils under a single new appellation: *Homo erectus.* The nomenclatural journey of the Trinil skull cap and thighbone had begun with *Anthropithecus erectus,* had become *Pithecanthropus erectus,* and finally became *Homo erectus,* the name by which the fossil is still known. It is considered by many to be a perfect missing link between earlier forms of *Homo* and later *Homo sapiens.* And by now there is sufficient evidence to convince many anthropologists that despite what Haeckel speculated about his hypothetical missing link, *Pithecanthropus alalus,* he could in fact talk.

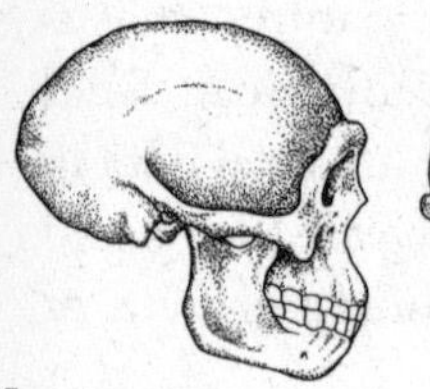
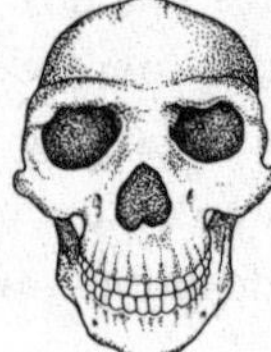

Homo erectus

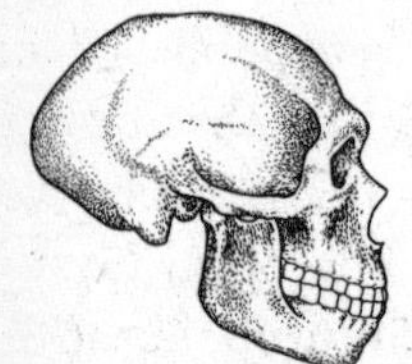
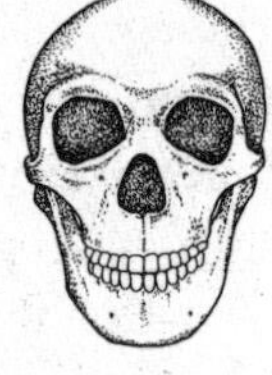

Homo neanderthalensis

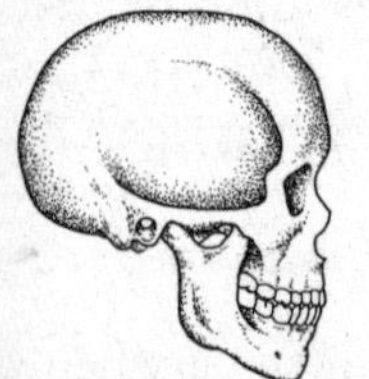
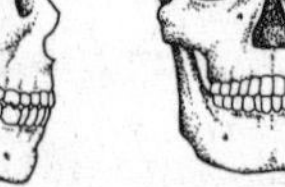

Homo sapiens

Weidenreich and von Koenigswald came to the conclusion that *Sinanthropus* and *Pithecanthropus* were the same creature, which came to be called *Homo erectus*, as shown in the top row. Notice that *Homo neanderthalensis* has some of the primitive features of *H. erectus*, such as the brow ridges and the low skull (it also has unique features, such as the protrusion of the mid-face), and some modern features of *Homo sapiens*, including a large brain.

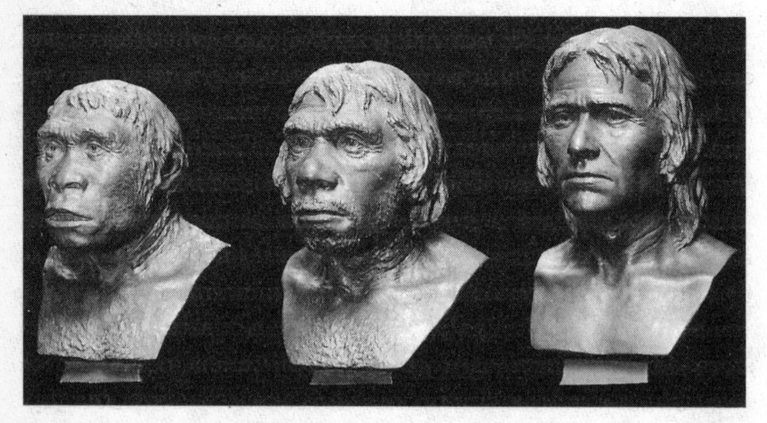

Early reconstruction of *Pithecanthropus* (left), Neanderthal (center), and *Homo sapiens*. (L. Boltin, American Museum of Natural History)

6

The Child Has a Date

WHEN we returned to Berkeley after our trip to Java in the fall of 1992, we were excited at the prospect of getting new dates for the pumice layer at the Mojokerto site, using the very precise laser-fusion argon-40/argon-39 technique. It promised to be an important event in the life of our geochronology group and the Institute of Human Origins, of which we were a part. What we didn't know, however, was how tumultuous would be the year and a half we were entering. "Manic-depressive" might be the appropriate adjective, as circumstances flung us between tantalizing promise and agonizing frustration, between elation and deep desperation. At one point the entire Java project seemed doomed, when it looked as if our professional base, the geochronology group at the institute, would cease to exist.

In 1989, the Berkeley Geochronology Center's initially loose relationship with Don Johanson's Institute of Human Origins had been made more formal, so that administratively the anthropologists and the geochronologists were one unit, with Johanson its president. From then on, the geochronology

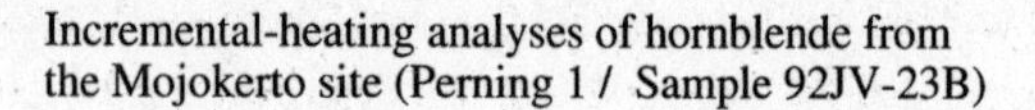

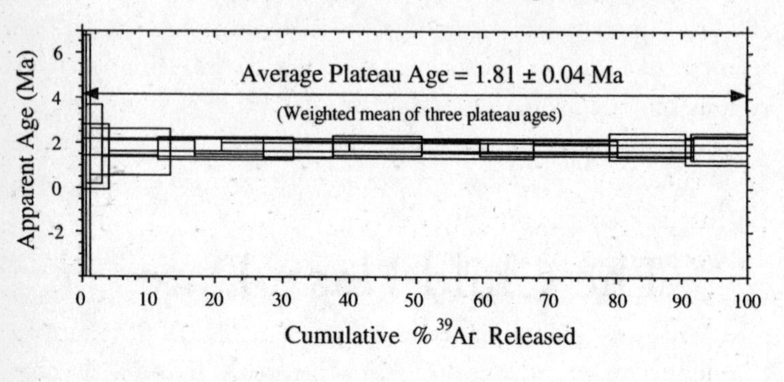

Inverse isochron of the same analyses

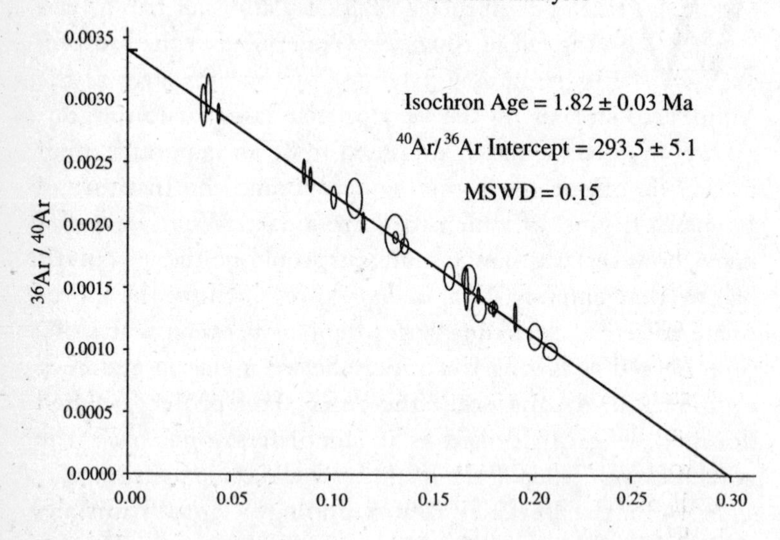

Data associated with obtaining an age of 1.81 million years for the Mojokerto child's skull.

group would be known as the Institute of Human Origins, Geochronology Center. Each November the institute's board gathered in New York for its annual meeting, and it was an opportunity to give brief reports of what was going on with various science projects. By the time of the 1992 board meeting, I had already obtained one dating result for the pumice from the Mojokerto site, close to 1.8 million years. "I was delighted," Garniss remembers, "because it vindicated my earlier work, and this time there was only a small margin of error, about 2 percent, as opposed to the 25 percent I'd gotten back then." Because it was just one analysis out of half a dozen we planned to do over the coming months, we didn't do a big presentation at the board meeting, but we did tell people about it. The result, though preliminary, generated a lot of cautious interest.

But in Bill Kimbel, the institute's director of paleoanthropology and a talented anthropologist, it generated a—to us—bizarre response. "Why didn't you ask one of us to come with you to Java?" he said to me. "You should have had an anthropologist with you!"

I was surprised and annoyed. "Don't you remember you and Don saying it was a waste of money when we asked for support from the contingency fund?" I snapped back. Bill and Don were certainly on the side of the skeptics concerning the possibility of sorting out the Javan dates, and had said so in no uncertain terms. "Don't you remember saying it was a bunch of crap and a waste of our time? In any case, we did have an anthropologist," I continued. "We had Jacob."

Bill didn't have an answer to that. Now that our project, which was once ridiculed, was a potential headline maker, it looked to us as if Don and Bill wanted to muscle in. Don had been deeply suspicious about our funding of the project in the first place. Apparently he had forgotten about the $6,000 grant from the institute's contingency fund, so he went to Gordon Getty to ask if Garniss had got the funding from him. Getty was a major financial supporter of the institute, to the tune of $1 million annually in matching funds, and Don didn't want anyone in the institute asking him independently for money. Don's accusation displeased Getty, who told him that Garniss had not asked for financial support for the trip. It has to be said that the relations between the anthropology and geochronology sides of the institute were not the best; such exchanges were typical and say much about the lack of mutual regard and trust.

In the months following the November 1992 board meeting, I continued to run analyses on pumice from the Mojokerto site, and continued to get a date close to 1.8 million years—1.81 million, plus or minus 40,000 years, to be precise, and in the world of geochronology that result *is* precise. We were in no hurry to rush into print with the result, for three reasons. First, no one else was working on this, so we didn't face the danger of being preempted. Second, if we published the Mojokerto date by itself, we faced the legitimate criticism that it was a freak result. We had also collected samples at Sangiran, the canal-side site of discovery of two pieces of skull, S27 and S31, which most people thought were 700,000 years old but we suspected were much older. If we were right—if we did get dates much older than 700,000 years for the Sangiran fossils—then it would bolster the credibility of an ancient age for Mojokerto. But the third reason was the most powerful, and a potential clincher in the decades-old bickering over the Mojokerto child's age: we felt we had an ace up our sleeves, and we wanted to play it.

Because the child's cranium is packed with pumice, rather than some other kind of rock, it offered the opportunity of obtaining an age that is virtually ironclad: date the pumice in the cranium and you can be pretty certain that this is the age of the fossil. Then the issue over exactly where the child's cranium had been found would melt away, because we wouldn't be relying on a date obtained from pumice out in the field; it would come right from inside the skull itself. There could be a question of whether the child's skull had become buried in an ancient ash layer, so that the age of the ash would be far older than the child, but there were no telltale signs of the rough-and-tumble journey this would have entailed, no extensive abrasion on the surface of the cranium, for instance. But because of Jacob's legendary protectiveness over the fossil, we were unsure whether he would agree to the removal of sufficient pumice from the skull for our needs, small though that amount was. Garniss composed a request in the most diplomatic language he could muster, and faxed it to Jacob in late December. Within two weeks Garniss had a reply from Jacob. He was enthusiastic about the project, he said, and would like to see the long-running question mark over the Mojokerto child's age removed forever.

This was in January 1993. Removal of pumice from the child's skull would be a momentous event in the fossil's history, so we planned to capture the whole event on film, which would mean some extra planning to get the equipment together. But we were pressed for time because Ramadan was coming up in not much more than a month; in observance of the religious festival, Jacob would not be in the museum for several weeks. No Jacob, no fossil. We realized we would have to go in February.

We were having lunch with Ann Getty, Gordon Getty's wife, one day in late January, talking about the Mojokerto dating results and the planned visit to Java. She became so excited that she wanted to go right then. Now! That very day! Ann is a very now kind of person, always pushing to get things moving, and she usually does. Ann was funding the trip, as with such short notice we would not be able to get money elsewhere. There simply hadn't been time to go through the necessary bureaucratic hoops, for one thing. We explained to Ann that it was impossible to leave right then, because we had some organizing to do. Ann, who is an enthusiastic and learning paleontologist, had to wait two days on this occasion before she got things moving.

In addition to removing pumice from inside the child's skull, we had two other objectives for the trip. One was to go back to Mojokerto and scour the section even more thoroughly than we had done on the previous visit, to see if there was a layer of pumice anywhere else, older or younger, that the skull could possibly have come from. The second was to collect more rock for a second type of dating analysis, known as paleomagnetic dating. Periodically in Earth history, and for reasons that are little understood, the Earth's magnetic field flips, so that a compass needle that normally points north, as it does now, would point south instead. The current direction of the magnetic field is called "normal," and the reverse is called "reversed." (Geologists will win no literary prizes when it comes to nomenclature!) During the formation of certain rocks, the prevailing direction of the Earth's magnetic field is recorded in iron-bearing minerals that act like miniature compasses. This residual magnetism can be detected with sensitive instruments, and by noting the magnetic polarity in a series of rocks of known ages, geologists have created a profile of the pattern of reversals through much of Earth history. In the past 3 million years, for instance, there have been at least ten such major events.

The paleomagnetic polarity time scale is helpful to geochronologists as supporting evidence only. Knowing that a particular rock is magnetized in the normal direction doesn't give you its age, because there have been at least six periods during the last 3 million years when the polarity was normal. But if, say, you obtain a date of 1.5 million years for a pumice sample by radiometric dating, and the magnetic polarity is normal, then something is probably wrong, because between 1.1 million and 1.8 million years ago the polarity was primarily reversed. In the case of the Mojokerto child, the age we had for it, 1.81 million years, indicated that it should fall in what is called the Olduvai Event, which displayed normal polarity between 1.77 million and 1.95 million years ago. We therefore planned to find the polarity of the layer of rock in which the child was said to have been found: if it was normal, then this would be consistent with an age of 1.81 million years; if it was reversed, then we would have a problem.

So, at 12:30 in the morning on 1 February, we flew out of San Francisco, bound for Java, aboard Singapore Airlines Flight 1. Also in the group were Ann Getty; her assistant, Meg Starr; and Susan Antón, then a graduate student in the anthropology department at U.C. Berkeley, studying under Clark Howell. We were in a mood of tense anticipation and high expectation.

A Minor Delay

We arrived in Yogyakarta the following day, in the early afternoon, exhausted. We checked into our hotel and planned to rest, but we called Jacob first, to tell him that we had arrived and to find out which day later in the week would be good for working on the skull.

"Come to the lab right away," Jacob said.

"Now?" Garniss replied, stunned. "But we've been flying all night!"

"Now; we have much to do," Jacob insisted.

Reluctantly we gathered our photographic equipment together and took a cab to Gadja Mada University, where we found Jacob contemplating the child's skull, which was resting on a table in mute welcome. A small hammer and chisel lay by its side.

We exchanged greetings with Jacob, who was unusually garrulous, and then he said, "Let's get to work." Garniss pleaded that we were all exhausted and would prefer to start the next day. "My schedule is very tight," Jacob explained. "We have to do it now."

In what felt like a drugged state, Susan and I began setting up for what would be a historic photo shoot, trying to focus our fogged minds and invigorate our enervated bodies. Jacob had apparently already been working on the matrix in the skull before we arrived, and had removed a tiny fragment, which he gave to Garniss.

"Can we have some more?" I asked.

"Of course," Jacob replied, in a surprisingly helpful mood. He picked up the hammer and chisel and was about to attack the matrix once again when I persuaded him to wait until we were ready with the cameras. Actually, I was alarmed by the whole situation. The skull was precariously poised on some books, and I was afraid Jacob might split the skull or send it rolling off the table.

Susan and I were so jet-lagged that it took us quite a while to set up the lights and tripod. Then Susan discovered that in our haste to leave Berkeley, the film I had brought didn't match the lights she had brought. We had to find either a blue filter or some tungsten film, or the pictures would come out yellow. Leaving the rest of the group in the lab with Jacob, I went searching through what seemed like every photo store in town. It took a while, but eventually I tracked down the required filter and returned to the university. By 4:30 everything was ready—except Susan and me. We were profoundly tired and weren't thinking straight. Afraid there was a good chance of our making a mistake while we were in this state, I suggested that we postpone the work until the next morning.

Jacob cheerily agreed. "Come at nine o'clock," he said. "We can do it then."

Much relieved, we all returned to our hotel, had a Bintang beer, a light dinner, and slept.

At 9:00 the following morning we turned up at Jacob's office, just as he had suggested—and found Agus Suprijo, Jacob's assistant, but no Jacob. "I'm afraid I have bad news," said Agus, obviously in great discomfort and embarrassment. "Professor Jacob is in Jakarta. He has a meeting."

We were dumbstruck. Garniss asked whether this was an unexpected trip.

Agus said it wasn't, that Jacob knew he was going. "I don't know why Professor Jacob didn't tell you," Agus continued, "but he said it was all right for you to go into the field to collect samples for the paleomagnetic work."

It was annoying and bewildering. If Jacob knew he was going to be in Jakarta this morning, why did he ask us to meet him in his lab? But this was Java, and things like that happen in Java. We said we'd be back in three days. Three days weren't going to make any difference in the long term, after all.

A Major Blow

We decided we would fly to Surabaya, which is close to Mojokerto, because it was too long a drive in the short time we had. The trip was uneventful, and we successfully collected the samples we needed for the paleomagnetic analysis. We did have a chance once again to look for pumice layers higher in the section. Like our previous attempt, we found none. The flight back was not so uneventful, unfortunately, as the plane left during a violent thunderstorm with terrific lightning all around us. The little plane bounced five times during takeoff, and it flew like a kangaroo. We really thought we weren't going to make it. We survived, however, and arrived in Yogyakarta late at night, and were eagerly anticipating the work we had principally come for.

When we walked into Jacob's lab the following morning, we were surprised to see that all our camera equipment had been moved, stacked up by the wall. "What's going on, Agus?" Garniss asked.

"Professor Jacob has left a note saying he is too busy to do this now, and he can't see you," Agus explained, again in obvious, excruciating embarrassment. He couldn't look Garniss in the eye. "Professor Jacob says that this is the end of the work," he finished, his voice trailing off to a whisper. We felt powerless and enraged.

We hung around the lab for a while, wondering just what game Jacob was up to, wondering what we should do next. Eventually a message came from Jacob that he would see Garniss. "He was very affable throughout our meeting," remembers Garniss, "asking whether we had a good trip to Mojokerto, saying that his flight from Jakarta had been awful, with a mother and child throwing up all over the place. He'd flown through the same storm that we had. I didn't say anything about our agreed-upon plan to remove pumice from the skull, didn't get angry. I knew that wouldn't get us anywhere, and I wanted to be able to come back another time. Eventually Jacob wished me a safe trip home, all smiles. And that was that."

Seething, I sat down and wrote a long letter by hand, expressing my disgust in distinctly intemperate language. The letter said something like "Why have you wasted our time like this?" and "If you don't want to work together, that's fine, but don't do this half-assed thing to us." Jacob never responded, but I wasn't really surprised.

We were devastated. How had such a thing happened? What was Jacob thinking? Wasn't he as eager as we were to solve the Mojokerto riddle once and for all? We had lots of questions but no answers. Ann wanted to rush off to the American Embassy in Jakarta and have someone do something about Jacob, but Garniss knew that wouldn't get us anywhere. Instead, we decided we would salve our collective wounds by spending a few days in Bali on the way back to the States.

We left that afternoon, taking with us the a piece of pumice that was too small to date, and a lot of "if only's." If only we brought the right film with us, we wouldn't have had to waste time searching for a filter and could have done the extraction and photography that first afternoon. If only we hadn't discouraged Jacob from chiseling more pumice from the skull when he had offered, then we would have had sufficient material for dating, even if we hadn't documented the event on film. If only we hadn't suggested postponing the work until the following day, we would probably have got our pumice and our film, risky though it would have been.

But if only's don't get science done.

Final Preparations

Back in Berkeley, trying to put the whole sorry incident behind us, we got on with the work we could do rather than mope over the work we had been prevented from doing. More analyses of pumice from the Mojokerto site gave results similar to the earlier ones, 1.81 million years. And the paleomagnetic work showed the section to be of normal polarity, which is consistent with that date. More important, however, was the date we got for the Sangiran fossil: 1.66 million, plus or minus 40,000 years. That is almost a million years older than most people believed. People might still question what we would say about the Mojokerto child and its age, but with Sangiran being similarly ancient, people would surely listen. People would have to accept the idea that humans moved out of Africa a whole lot earlier than everyone believed. We began to draft an article that we planned to submit to the journal *Science*.

By this time we were becoming ever more strongly convinced that the site Jacob had taken us to in 1992 was indeed the place where Andojo had found the child's skull six decades before. (Four subsequent visits, on which we were accompanied by Garry Scott and Joe Butterworth, two geologist colleagues, strengthened that conviction, through extensive examination of the geology and studying the geological maps made by J. Duyfjes in the 1930s and 1940s.) Because the cranium is filled with pumice, it must have been buried in a pumice layer. When we revisited the site during that ill-fated trip in February 1993, we looked once again for pumice higher in the section than the pumice layer we had analyzed. There was none. And below the pumice layer we saw that you'd soon get into marine deposits, so the cranium couldn't have come from lower, either. The pumice layer we had dated *must* be the right one.

We were able to bolster this contention with the meager piece of pumice that Jacob had given us in February. Insufficient as it was for dating, it nevertheless was enough for microprobe analysis, a technique that produces a profile of all the chemicals in a rock, a kind of chemical fingerprint. I analyzed minerals from the pumice layer at the Mojokerto site and the pumice from the child's skull. The profiles of the quantities of eleven chemicals in the two samples were very, very similar. That finding didn't confirm that the pumice from the Mojokerto site and from the child's cranium were one and the same, but at least it was consistent with that possibility.

We now felt we were prepared for the hail of skepticism that we expected would greet the publication of our *Science* paper.

Big Headlines, Heavy Skepticism

We wrapped up the writing of the *Science*[1] paper by the fall of 1993, gave a long presentation of it to the Institute of Human Origins board at their meeting in early November in New York, and submitted it to the journal that month. Papers submitted to *Science* often take a long time to appear in print, but the editors said they would push this one through quickly so that its publication would coincide with the annual meeting of the American Association for the Advancement of Science, to be held in San Francisco in late February 1994. As is typical of the annual AAAS gathering, the San Francisco event was a vast marketplace of new research, with hundreds of scientists presenting their latest work to an audience of several thousand. It's always a popular venue for journalists—there were more than six hundred of them at the 1994 meeting—because of the range of science on display, from climate change to AIDS research, from neuroscience to cosmology. Garniss was to give a short presentation about our work in Java and had been told there would be a press conference afterward, which often happens with topics the AAAS believes will be popular.

Imagine our reaction when we walked into the room for which the press conference had been assigned. Empty. Not a reporter or a notepad in sight. Just a few empty chairs.

"I guess our story isn't so popular after all," I said to Garniss.

"I guess not," Garniss replied. "Let's just head back to Berkeley."

As we were leaving, one of the organizers found us and said that the press conference had been moved to another room. That room—a much bigger one—turned out to be packed. All the chairs were taken, and many reporters stood at the back. We made our way to the front, where half a dozen microphones sprouted from a long table. I was thinking, "Oh my God, what *is* all this?" Neither Garniss nor I had done any real preparation for a crowd of that size, and we felt oddly naked. Clinging to our only prop, a couple of casts of the Javan fossils, for protection, we each gave a short talk, Garniss's about his earlier experience in Java and mine about the new results. And then came the barrage of questions, good questions mostly. These people knew our stuff.

We expected a decent amount of coverage in the newspapers after that event, but we still weren't prepared for the volume. The AAAS people later told us that stories had appeared in 221 newspapers in the United States alone, reaching almost 24 million readers. Ours was the fifth most popular presentation of the meeting, judged on the volume of readership for the stories. We even made the front page of the *New York Times,* under the headline "Asian Fossil Prompts New Ideas on Evolution." The article, by veteran reporter Malcolm Browne, stated, "New evidence strongly suggests that the human race migrated from its African cradle nearly one million years earlier than most scientists believed. The discovery will force anthropologists to reexamine long-held beliefs about how and where evolutionary turning points made mankind master of the world."[2] Exactly right. One of our local papers, the *San Francisco Chronicle,* also had a front-page story, headed "New Fossil Date Shakes Up Ideas on Evolution." The reporter, Charles Petit, with whom we had talked the previous week, wrote that the new date suggested that "mankind's evolution into modern *Homo sapiens* is a far more complex story, on a much vaster stage, than textbooks now say."[3] Exactly right, too.

The newspaper reports came out right away and were mostly a straight presentation of what we had said at the press conference. For longer, more in-depth stories that contained other researchers' reactions to our work, we had to wait for the news magazines. *Discover, New Scientist,* and even *Time* magazine all had cover stories prompted by our work. We were gratified to read in *Time* that Clark Howell, an anthropologist at Berkeley and a board member of the Institute of Human Origins, said, "This is just overwhelming. No one expected such an age."[4] Tim White, an anthropologist at Berkeley, brought up what we knew would be voiced: the supposed uncertainty of the Mojokerto skull's site of discovery. "The dates are fine," he told the reporter for *New Scientist.* "That's not an issue. The problem is [that] because it was discovered sixty years ago, how can you possibly be sure you're sampling minerals from the right site?"[5] Milford Wolpoff, of the University of Michigan and one of the more colorful of contemporary anthropologists, was even more direct. "I'm skeptical of anything in Java," he told *New Scientist.* "I'm not convinced that it's possible to relocate the site where the child's skull came from."[6]

In a news article in *Science,*[7] Frank Brown of the University of Utah and one of the best field geologists there is, voiced the possibility that the skull might have come to rest in sediments much older than itself, as sometimes happens when sediments get churned up, perhaps through water erosion, and redeposited. But that process causes the once discrete layers of sediment to become mixed up, with older rock often higher in the section than younger rock. That was not what we'd seen at the Mojokerto site. The section was chronologically consistent, with the age increasing as we looked from top to bottom, from younger to older sediments.

Not long after our *Science* paper was published, Frank visited our geochronology lab as part of a contingent from the Leakey Foundation. He told us that he'd made his remarks to the reporter for *Science* without having seen our paper. We described the section to him, explaining that there was no sign of reworking, or of channels cut into the sediment, or of anything that would support the possibility he'd raised. He seemed pretty much convinced. In any case, we felt we'd done as much as we could to counter the skeptics, but there would always be some of them.

Support from the Past, Too Late

A few weeks after the media fuss had died down, Susan Antón came to me and said, "Look at this! You're not going to believe it!" She handed me a photocopy of an old paper by Helmut de Terra, entitled "Pleistocene Geology and Early Man in Java." (We mentioned this paper in chapter 2.) Published in 1942, it had escaped our notice completely until this point. De Terra had visited the discovery site of the Mojokerto child in April 1938, with Harvard University's Hallum Movius and the French philosopher and paleontologist Pierre Teillhard de Chardin,

The Mojokerto site, taken by Helmut de Terra in 1938, showing von Koenigswald in a pith helmet. (Copyright by the President and Fellows of Harvard College, Peabody Museum, Harvard University)

under the guidance of Ralph von Koenigswald. That was just two years after the child's skull had been discovered, so the excavation site must have still been fresh in people's minds. Indeed, de Terra wrote that "the excavation pit in which the skull had been found looked as fresh as though it had been dug only a few months prior to our visit."[8]

There was a photograph of the site at the end of the paper, but it was so grainy that admittedly we could not tell for sure whether it was the site we visited. But de Terra described the site in some detail in the body of the text, and it sounded right to me. When Susan Antón and I read the description, our reaction was immediate: "Oh my God. It's the same thing we saw." De Terra had also done what we did—that is, he had checked the pumice in the skull to see if it matched that at the site. "What is of decisive importance for establishing the age of the Mojokerto fossil is the fact that the matrix adhering to the original skull can be matched with that from the layer exposed in the pit . . . This convinced me of the Early Pleistocene age of the fossil," he wrote.[9] His estimate of age was based on the types of animals that were living contemporaneously with the human skull. Early Pleistocene encompasses 1.81 million years very nicely. In a paper he wrote for Harvard's Peabody Museum, Movius said much the same thing. "Not only does the breccia inside the skull match the material exposed in the pit, but there is no other horizon in this vicinity from which it could have been derived," he observed. "Thus *Homo modjokertensis* [the Mojokerto child] is the oldest human fossil from the continent of Asia."[10]

When de Terra saw the child's cranium, he noted that the pumice it contained was white. Why and when it was painted black is a mystery to us and to Jacob. According to Phillip Tobias, a major figure in paleoanthropology for many years at the University of the Witwatersrand in Johannesburg, it was meant as a

Unpublished photograph of the Mojokerto site, taken by von Koenigswald in 1936 or 1937. The figure by the banana tree is possibly the discoverer of the child's skull, Andojo. (Copyright, Senckenberg Museum)

disguise of sorts. "Oh, I know why they painted it black," Tobias told Garniss a few years back. "It was to keep the Japanese from identifying it. Von Koenigswald told me that." Oddly, though, von Koenigswald apparently never told Jacob, with whom he had extensive professional contact.

We were pleased we had come across de Terra's paper because it confirmed what we knew about the site, but we felt stupid we hadn't known about it earlier; if we had, we could have cited it in our *Science* paper. But we were in good company: apparently no one else knew about it, either. You never see it referenced in the literature.

What you read in the literature are bold statements to the

effect that "there is no contemporary documentation of the Mojokerto discovery site, so it will be difficult if not impossible ever to relocate it," period. To us, the presence of the 1938 photograph in de Terra's paper opened the tantalizing possibility that other, similar, shots may be lying unseen in an archive somewhere. After all, three people had been on that 1938 visit: in addition to de Terra, there was Hallum Movius and Teihard de Chardin. Moreover, von Koenigswald probably had had several opportunities to take pictures. It seemed unlikely to us that they would have taken just one photograph among them. Thus we began a search across three continents that lasted six years and culminated with a revisit of the site in February 2000.

We started in 1994 in Bandung and Yogyakarta, reasoning that perhaps when von Koenigswald returned some of the fossils to Java, he might have included some archival material. If in fact he had done so, no one we spoke to knew anything about it. A dead end.

The following July Susan Antón and I went to the Senckenberg Museum in Frankfurt, Germany, where Susan planned to study some of the Sangiran fossils that are still stored there. We found one photograph in von Koenigswald's archives that looked like the one de Terra published in his 1943 paper, but unlike the published, close-up photo, this was one was taken some 40 to 50 meters away. In this photo, there is an Indonesian man standing next to a white pith helmet (the type von Koenigswald was wearing in the close-up photo), marking what appeared to be the same site shown in the published close-up photo. It was clear from the vegetation that this was an earlier photo, and, if von Koenigswald was the photographer, given the presence of his hat on the site, then the man in the photo could be the fossil's discoverer, Andojo. We know that Andojo had taken von Koenigswald to the site soon after he had discovered the child's skull, so it was a plausible inference.

Unfortunately, the photograph wasn't labeled, so our line of inference and connections between the photos would remain questionable to those who doubted the location to begin with.

Susan and I also tried a few other leads, by sending enquiries to various archives in the US, such as the American Museum of Natural History, where Franz Weidenreich and von Koenigswald studied the Java fossils in the late 1940's following World War II, and the Philadelphia Academy of Sciences, which had sponsored the 1938 expedition. We drew a blank in Philadelphia.

In 1996, while Susan and I were visiting the American Museum in New York, we had a chance to look through some of Weidenreich's archives. Accompanying some notes Weidenreich had made on the Java fossils, we found one grainy photograph, similar to the one we'd seen in the Senckenberg Museum, but it had been taken from a different angle and at a different time. This photo was labeled "Find Site of *Homo modjokertensis.*"

At about the same time in our search, we also considered looking at the Hallum Movius archives at the Peabody Museum at Harvard. Since Movius was part of the 1938 expedition, perhaps he had taken a photo of the site? We talked with Ofer BarYosef, an archeologist at the Peabody, and asked about Movius's archives. The news was discouraging. "The archives are in disarray," Ofer said. "There is a lot of stuff in boxes, papers and photographs, but nothing has been sorted. It would take you months to go through it." We felt we had to give up on that lead for now at least.

In mid-November 1999, I was taking part in a conference on Olduvai Gorge, organized by Rob Blumenshine of Rutgers University, sponsored and held at the Peabody Museum. I had started to do some geological work at the Gorge a little more than a year earlier, so Rob had included me in the gathering.

Otherwise, there would have been no reason for me to be there. On the last day of the conference Susan joined me at the Museum for a short visit with Ofer. We asked Ofer again about the Movius archives, more out of curiosity than any specific expectation. "Funny you should bring that up," Ofer told me. "The museum has spent the last two years sorting out Movius's archives, and they are now fully catalogued and documented. No one has looked through them yet." Susan and I immediately thought to look at them the next day, but it turned out that the Museum would be closed. Some kind of holiday.

When we got back to Berkeley I sent an e-mail to Sarah Demb, the museum's archivist, asking if she knew whether Movius's archives contained any materials on Java, primarily from the 1930s. She responded with a two-page list of items labeled Java correspondence, notes, and photos, although nothing specific about Mojokerto, but sufficiently encouraging to warrant a trip back East, which we did on December 13. We felt confident that we might be close to getting what we needed, but were crushed when Movius's voluminous material yielded nothing of the Mojokerto site. Just when it seemed that we were going to be cheated in our quest yet again, Susan said, "What about de Terra? We know that there was a box of correspondence between him and Movius, so there might be photographic archives as well." Nasrin Rohani, the photo archivist, quickly retrieved the box for us. In it, Susan and I found a series of photographs and, more surprisingly, a box of negatives with a note from de Terra to Movius saying, "I'm not working in Java anymore, and I thought these might be of use to you."

Very soon we unearthed not only the original photograph in de Terra's 1943 paper, but also five out of a series of five photographs and negatives of the Mojokerto site, taken from various angles and distances, all taken in 1938. We now had seven photographs, the Senckenberg photo probably taken in 1936, the American Museum photo from circa 1937, and five in 1938, more than any of us had dared hope for. The more we looked at our trove, however, the more it became clear that what we could see was not the steep gully that Jacob had taken us to in 1992. It was probably close, but it wasn't the place where Jacob had placed his hand on the wall of the gully and said, "This is where the fossil came from." We would have to go back Java and try to match the topography of six decades ago as depicted on the photographs with how it stands today. We knew it wouldn't be easy, both because the vegetation has changed, but also because of extensive farming and terracing that has taken place during that time. But we didn't know just how difficult it would be.

A retirement ceremony and associated scientific conference for Teuku Jacob early in February 2000 in Yogyakarta provided a timely opportunity to revisit Mojokerto. We had been in Java at that time of year on previous occasions, and knew it would probably be wet. But not as wet as it turned out to be. Susan, Garniss, Agus, and I flew to Surabaya early on the morning of the tenth, rented a Kijang, and arrived at the site at about nine in the morning. Although the weather looked promising, as we reached the village of Perning it began to rain. We stopped and bought a couple of straw hats to keep us dry and proceeded up the road to the Mojokerto site. We split up and searched in different directions, trying to match the angles from which the different photographs had been taken. The conditions were about as bad as they could be as we tried to make our way through mud that threatened to tear off our boots, and negotiated streams that are usually dry gullies. Moreover, the vegetation was so lush that it often obscured views we badly needed. By the end of the day, however, we could see that the discovery site lay somewhere on the ridge above the steep gully. It must be within a few meters, or perhaps a few tens of meters, distant

from the Mojokerto monument, which itself was obscured with vegetation and in a state of disrepair.

Although we are certain that the photos had to have been taken in the general vicinity of the monument, we couldn't pin point the precise spot of the site shown in the photos. We would have to try again in the dry season. From a geological and geochronological point of view, we can say that the trail on which we had been led by the newly discovered photographs support the fossil's age as we had reported it in the *Science* paper. Because of the low dipping angle of the strata, you have to travel about 100 meters north of the monument before that age would no longer be appropriate, and that is out of the question given what we saw on our visit, hampered though we were. At least as important, however, is that a site that has been widely assumed to have no contemporary photographic documentation turns out to be among the best documented of its age.

An Old Archeological Puzzle Solved

If we are right about the Javan dates, there are three major implications for the way we understand human prehistory. The first, and perhaps the most profound, concerns how and when modern humans, people like us, evolved. Second, we gain a different and deeper understanding of what kind of creature *Homo erectus* was. Third, we can suggest an explanation of a long-standing puzzle in archeology, which has to do with the role of technology in the dispersal of humans beyond Africa. The first two issues deserve chapters of their own, which will come later. The third will be addressed here.

The earliest known *Homo erectus*-like creature in Africa was found in Kenya and is close to 1.8 million years old, very similar to the date we obtained for the Mojokerto child. When anthropologists believed (as many still do) that humans first ventured

into Asia close to a million years ago, we faced a conundrum. Why did *Homo erectus* people wait for almost a million years before extending their range beyond Africa? And what was it that enabled them to do so when they did? The most popular answer was simple: technology. Early humans started to fashion stone tools around 2.5 million years ago, as archeological sites of that age in Kenya and Ethiopia attest. These tools were extremely simple, being little more than apple-sized cobbles of lava from which a few flakes were removed. These so-called core tools were probably used for chopping, scraping, and rough cutting. The small, sharp flakes themselves were also used as tools, as slicing implements. Together, core tools and flakes constitute what archeologists call the Oldowan technology, after Olduvai Gorge, where Mary Leakey systematically characterized them through decades of painstaking study.

Although making these tools is beyond the manipulative and cognitive capacities of modern apes, as experiments with a pygmy chimpanzee have demonstrated,[11] archeologists and psychologists believe that their manufacture required only limited mental abilities and did not involve the toolmaker's having an image, or mental template, of what shape was to be produced. About 1.5 million years ago, however, a new, more complex kind of toolkit began to be produced in Africa, called the Acheulean by archeologists, after St. Acheul, in France, where they were first found. The most striking implement in the assemblage is a teardrop-shaped hand axe, often the size of an open hand. There's little question that the makers of these tools had a mental template when they were making Acheulean axes, which required much more skill to make than Oldowan tools. The toolmakers had to remove many flakes, and in a systematic fashion, in shaping such axes, some of which are aesthetically beautiful. And the process took at least twenty minutes, as judged by the experience of archeologists

who make them today. The axes are also much more versatile and powerful implements. They can cut through thick, tough hide, for instance, which may have enabled the toolmakers to include much more meat in their diet than was possible previously.

Acheulean hand axes became emblematic of a new way of life for *Homo erectus,* who is assumed to have been the toolmaker—a way of life that was much more like that of modern hunter-gatherers than was possible in earlier times by earlier people. The ability to make and use hand axes, suggested anthropologists, is what enabled *Homo erectus* to live in lands they were unable to live in previously, making possible the journey from Africa to Asia. This move probably should not be envisaged as a purposeful migration, with people setting out on a journey with a destination in mind; it was more likely a gradual expansion of their range. The *Homo erectus* people took their newfound technology with them, as is indicated by the many caches of Acheulean axes in Eurasia.

But therein lies another puzzle. Although these tools are found in Europe and western Asia, they are absent in eastern Asia, where chopping tools, like the Oldowan technology, are found instead. The division of the Old World on the basis of the presence and absence of Acheulean axes is known as the Movius line, after the Harvard archeologist Hallum Movius, who first described the pattern in the 1940s. Archeologists have conjured up several lines of argument to explain the pattern. Perhaps there was no suitable raw material for manufacturing these larger stone implements, for instance? This seems a little far-fetched to us. Or perhaps the residents of eastern Asia preferred a different type of implement, made from a different type of raw material? Geoffrey Pope, an archeologist at William Paterson College in New Jersey, argues that bamboo provides such an alternative. Ubiquitous in eastern Asia,

bamboo can indeed be fashioned into simple, effective knives, as some technologically simple people in the region do today. Javan *Homo erectus* may therefore have used small stone chopping and slicing tools to make bamboo knives, suggests Pope. Perhaps. But unlikely.

The new Javan dates offer a different solution to this puzzle, as well as to the conundrum of why *Homo erectus* waited so long—almost a million years—before going beyond the confines of Africa. Obviously, *Homo erectus* didn't wait. That geographic expansion began as soon as the species evolved, a little earlier than 1.8 million years ago. They *didn't* require a new technology to get to Asia, because no new technology was available back then. In retrospect, the technology-explanation scenario looks like special pleading, since many species of large mammal, such as elephants and antelopes, moved between Africa and the rest of the Old World in ancient times without benefit of technology. The fact that *Homo erectus* was Asia-bound as soon as it appeared has an important implication: it was a very different kind of animal from its forebears, one that was physically, and perhaps mentally, better equipped to move over large distances and live in novel environments.

Our new Javan dates also offer an explanation for the absence of Acheulean axes in eastern Asia. When people first arrived in Java, Acheulean axes had yet to be invented. Perhaps, unlike their relatives back in Africa, *Homo erectus* populations in eastern Asia never hit on the idea, continuing instead to work with Oldowan-like tools, which can be very versatile in the right hands. Populations that moved from Africa after the Acheulean industry had been developed apparently never settled in eastern Asia, prevented perhaps by populations of people already firmly established there. This explanation, more parsimonious than the others, seems to us very plausible. (Parsimony in science refers to simplicity of assumptions. One

solution is more parsimonious than another if it requires fewer logical steps, fewer leaps of the imagination.)

We said earlier that the oldest *Homo erectus*-like creature in Africa dates back to around 1.8 million years, a virtual contemporary of the Mojokerto child. On the face of it, therefore, it looks as if the journey from Africa to Asia was taken at breakneck speed. A simple calculation shows otherwise. If people moved at a modest 10 miles a generation, the 6,000-mile trek from East Africa to Java would have taken 15,000 years, a mere eyeblink in terms of human history. What we see as lightning speed would not have seemed speedy to the trekkers.

There is, of course, an alternative explanation of the virtually simultaneous appearance of *Homo erectus* in Africa and Asia, one that more than a few anthropologists favor and which we mentioned in our paper in *Science.* Perhaps *Homo erectus* evolved in Asia, not in Africa after all. The theory cannot be disproved at present, but so far there is no solid evidence for the presence of ancestors of *Homo erectus* outside the African continent. This could change tomorrow, of course, with the spectacular discovery of an *erectus* progenitor in Asia. (Garniss would love to see that happen, because it would shake everything up, big time.) Until that happens, however, the most parsimonious position is that *Homo erectus* was born in Africa, cradle of humankind.

Despite the gut-wrenching fiasco of the February 1993 visit to Yogyakarta, we felt we were still on track with making anthropological history by solving the long-standing problem of the Mojokerto child and its age. The wide public interest in our *Science* paper, published exactly a year later, and the professional acknowledgment of our work (although some of that was begrudging), had buoyed our spirits tremendously. We still had more work to do in Java, and we even nursed the hope that one day we might succeed in getting sufficient pumice to test from the child's cranium, with or without Jacob's help. We were proud of the work we were doing at the Geochronology Center—of which the Javan adventure was just a part—and we looked forward to more years of work in Java.

But, as the old saying goes, pride comes before a fall, and within weeks of the publication of the *Science* paper and the ego-boosting hoopla that surrounded it, we found ourselves pitched into a maelstrom of professional jealousy and infighting that threatened to put an end to the Geochronology Center at the Institute of Human Origins and to our quest for history making.

7

Rocky Marriage, Painful Separation

THE afternoon of 3 May 1994 was bizarre for Paul Renne, director of the geochronology division of the Institute of Human Origins. Three o'clock in the afternoon, to be precise. For the previous two hours, the institute's board had hauled itself through an excoriating meeting that board members later described variously as "terrible," "heart-breaking," and "very sad." The upshot of the event, which was decided by a voice vote of nine to four, was that our geochronology presence at the institute was finished. Effective immediately. At the stroke of three o'clock.

"There I was," remembers Paul, "with the awful prospect of going downstairs within a few minutes and having to tell my guys that they were out of a job, that they were no longer going to be able to do their work; and I was calmly shaking hands with the board members, these people who had just put me out of a job, too, exchanging pleasantries. You know, things like 'Nice to see you again,' and so on. I was in a state of denial, because as far as I could see, it just shouldn't have happened as it did."

Gordon Getty, an IHO board member and the institute's single most important financial backer, had called the emergency meeting, which was supposed to be followed the next day with a presentation of the institute's achievements to the World Presidents' Council, at the Claremont Hotel in Berkeley. A scientific celebration of sorts. And there was a lot to celebrate. In addition to our headline-grabbing dates for the Mojokerto child, announced in February and featured in a cover story in *Time* magazine, a month later the anthropology side of the institute could boast the discovery in Ethiopia of the first nearly complete skull of a 3-million-year-old human ancestor, *Australopithecus afarensis,* the species to which the famous Lucy skeleton belongs. Either achievement separately would normally have been sufficient for the thirteen-year-old Institute of Human Origins to count the year a success. Scientific accomplishments aside, however, Gordon had become increasingly disturbed with the way the institute was being run in general, and distinctly unhappy with the behavior of Don Johanson, its president, in particular.

Gordon, a man who likes things to go his way, wanted changes. But this time he got more than he asked for or expected. At the board's behest, and contrary to Gordon's proposal, the institute had ripped off its geochronology arm and tossed it aside, with the explanation that this was the only way the IHO could save itself.

With mute disbelief and then rising outrage was how the geochronologists greeted the news when Paul and Garniss descended to the bowels of IHO, where the geochronology laboratories were housed, and relayed the afternoon's events. None of us could understand why the board would do such a thing. It was literally incredible. But events continued to move quickly, giving us little time to articulate our collective anger and thoughts. A memo arrived from Susan Shea, the institute's executive director, within minutes, which demonstrated either Shea's speed and proficiency in typing and photocopying or her prescient anticipation of the meeting's outcome. The memo said, in effect, that as of that day geochronology personnel were no longer employed by IHO; that they would receive payment for the following day only; that people's possessions must be removed immediately from their offices; and that issues of liability prevented geochronology personnel from having access to their laboratory. She also announced that a meeting was to be held immediately in the library.

"The atmosphere was extremely strained," Paul remembers of the library gathering, "and strange." In a quintessential public relations voice and manner—that is, curiously detached from people as people and their realities—Shea announced that because of extreme financial crisis the board had decided to terminate geochronology activities, and that this action was very regrettable because it meant saying goodbye to friends and colleagues. Paul pointed out that the geochronology group had good grant support, including five recently announced National Science Foundation grants amounting to more than $700,000 over three years. Perhaps, he suggested, the decision to terminate could be put off while other funds were sought. To no avail. Shea, who had been in her job just four months when this storm broke, responded flatly but firmly: A quick, clean break was what the board wanted. No amount of argument to the effect that many important experiments were under way, and would be lost through this action, made any difference.

There was no discussion about transition or what we would do with our experiments. We were terminated, period. Anger and frustration were palpable. And more. Garniss, who, with Paul, was a member of the IHO board, blew up volcanically, as he sometimes does when pushed too far, and stormed out of the room. Johanson, who was sitting in a corner of the room wearing what I saw as a smirk on his face, mocked Garniss's dramatic departure. Angered by this mockery and by an additional muttered remark, I delivered an arm gesture in Johanson's direction, its unmistakable meaning internationally known.

A marriage of convenience of almost a decade's duration between the anthropologists and geochronologists who made up the institute was at an end. The fact of the breakup was no great surprise, because in truth it had been a rocky marriage for some while. But the way it ended had not been expected, sudden and traumatic as it was. And as too often happens with divorce, the sequelae of separation would be at least as painful as the conditions that sundered the union in the first place.

One likely casualty among many in this rising turmoil was an international conference of geochronologists that was to be held in Berkeley and chaired by Garniss, scheduled for early June. Another, it seemed to us, was any prospect of our continuing our work in Java, which we were due to visit again four short months hence. The one best chance we had of securely resolving the decades-long mystery of when humans first arrived in Java looked as if it would be scuppered by that nine-to-four voice vote on that May afternoon.

A Shotgun Marriage

Tim White once described his erstwhile friend and collaborator Donald Johanson as having the appearance of "a nail-polish salesman in Yves St. Laurent pants and Gucci sneakers."[1] While Tim, an anthropologist at Berkeley, is widely known in the profession for his sharp intellect and equally sharp tongue, Johanson is better known for style than for substance. Like his archrival Richard Leakey, Johanson pays great attention to his physical presence in the world, which, combined with great charm, creates a strong, positive impression, whether on the public lecture circuit or while fundraising among the social elite. And not only does Johanson share with Leakey an envied knack for being in the right place at the right time for making wonderful fossil discoveries, he also has a great talent for promoting the science of paleoanthropology in the public arena. As Gordon Getty notes, "Don has done for anthropology what Carl Sagan did for cosmology."[2] All sniping aside, that's important.

In the science of paleoanthropology, a single fossil is often sufficient to propel its discoverer to worldwide fame, providing that that person has the social dexterity to parlay to good effect what chance offers. *Zinjanthropus* was Louis Leakey's springboard to celebrity, for instance. (It was his wife, Mary, who actually found the fossil, of course, but Louis was the showman.) And the skull known simply as 1470 was Richard Leakey's passport to fame, in 1972. Both these finds were of great significance to the science: *Zinjanthropus* was the first early human fossil found in East Africa; and 1470 was the earliest known large-brained specimen of our own genus, *Homo.* But Johanson's ticket to this exclusive club was the most spectacular of all: the half-complete skeleton of a 3-million-year-old individual of the species *Australopithecus afarensis,* better known to the world as Lucy. The ink on his doctoral diploma was barely dry when Johanson announced the November 1974 discovery. There was no doubt that the young researcher's professional star was rising fast and rising high.

The Hadar region of Ethiopia, where Johanson came across Lucy's petrified remains, proved to be a treasure trove of early human fossils, and their discovery and subsequent interpretation had a tremendous impact on the science. Nothing as old or as primitive looking had been found previously in the annals of human prehistory. And when in January 1979 Johanson and Tim White published their thoughts on the place of their newly named species, *A. afarensis,* in the human evolutionary journey, it truly was a case of having to toss aside all previous evolutionary trees and craft new ones.[3] Lucy and her fellow members of *A. afarensis* were said to be the fount of all later humanity, of the many subsequent species that went extinct and of the one that survived, ourselves. Not everyone agreed with this view, however; most prominent among the dissenters was Richard Leakey.[4] And later discoveries—some made just recently—have shown that the evolutionary picture is more complex than Johanson and White could have envisaged at that time. But in many of their claims the two young anthropologists proved correct.

White was the principal analytical mind behind this work, but it was Johanson's name that the public knew and recognized. While White went on to hew a distinguished academic career at Berkeley, academia was too narrow, too constraining for Johanson. He had a dream, he said, "to create a research institution devoted solely to the study of prehistory."[5] As a student, Johanson had seen Louis Leakey in performance mode, pacing back and forth across the lecture stage, arms waving in expansive gestures, voice booming, his enthusiasm gripping a rapt audience. The young student, too, was enraptured, and the experience sowed a seed of possibility: that perhaps he, too, would one day speak to enraptured audiences about their origins, about the evolutionary odyssey of *Homo sapiens.* The institution that the young Johanson dreamed of would be the vehicle for both the study of human prehistory and its promulgation to a romance-hungry public.

In 1981 Johanson courageously left a salaried post at the Cleveland Museum of Natural History and moved to Berkeley, where he established the Institute of Human Origins in a small basement near campus. Although the institute was independent of the university, Johanson was enthusiastic about possible synergistic relationships with scholars in anthropology, such as Clark Howell (who had been Johanson's doctoral advisor at Chicago) and of course Tim White, and with the archeologist Desmond Clark. With independence comes freedom but also the constantly gnawing need to raise funds. Johanson was already on good terms with Gordon Getty (Johanson stayed at the Gettys' San Francisco house when he first moved to the Bay Area), and Getty soon became a financial supporter of the nascent institute. So too did the industrialist David Koch and the Ligabue Research and Study Center in Italy.

Within two years IHO moved to larger premises, sharing a building with, ironically, the Church Divinity School of the Pacific, an Episcopal seminary, again near campus. The institute gradually began to establish itself as a serious center for the study of prehistory, with a small full-time staff and visiting scholars from other institutions. From its inception, the institute never lacked for notoriety, brought both by a steady stream of important fossil discoveries in Ethiopia and Tanzania and, it has to be said, by an almost equal stream of controversy, which included more than one instance of its researchers being denied access to fossil sites under cloudy political circumstances.

No one doubted the seriousness of Johanson's venture, but clearly it was not living up to the scale of the original dream, partly because of its diminutive size and the paucity of full-time staff of international repute. When Bill Kimbel, a well-respected anthropologist, joined the institute from the Cleveland Museum of Natural History in 1985, the quality issue improved, but the size problem remained. With so small a scientific payroll, IHO could not be described as a center of excellence in the conventional scientific sense.

Meanwhile, Garniss was contemplating the end of a long, distinguished career in geochronology at Berkeley. His retirement was set for 1989, and his lab—much equipment and a group of followers who called themselves the Berkeley Geochronology Center—faced disbandment. "I had started to think about all this about four years earlier, in 1985," recalls Garniss. "It was forced on me, really. I knew I didn't want to give up the work, but the idea of establishing a new lab outside the university by myself seemed difficult at best." One day Garniss was having a conversation with Clark Howell, who had been on the board of IHO from the beginning, and was musing about the future and its uncertainty. Clark suggested that Garniss might contemplate joining forces with Johanson at IHO. "It was a natural idea," Clark now says. "IHO only had two scientists [Johanson and Kimbel], and one of them was really a PR man. Bringing the paleoanthropology and geochronology groups together would give the place some scientific depth."[6] Garniss was enthusiastic about the prospect, and Clark promised to bring a proposal to IHO's board.

Although Clark considered the proposed union "a natural idea," he had no illusions about how life might be at the enlarged IHO. Clark, a man of encyclopedic memory where his science is concerned, knows everyone in the business very well, too, and knows what can happen when egos clash. "I said to Garniss that he would have to 'take Don for who he is,'" Clark recalls. "And I told Don that 'Garniss is not easy to live with . . . He's demanding and getting on in years.' I knew this venture was important for Garniss, and I wanted it to work. But it was a shotgun marriage, not a marriage made in heaven."

In the fall of 1985, Garniss, Bob Drake, Alan Deino, and some of us who were still at the time graduate students started to install ourselves in the newly renovated basement of IHO's quarters. We continued with our previous name, the Berkeley

Geochronology Center, and had at first what was a rather loose association with IHO. Although we carried the greater scientific clout in terms of number of researchers, our relationship with the anthropologists of IHO was like a bunch of bright but unruly kids being taken under the wing of a caring relative. While IHO was an established institution, legally and administratively coherent, the BGC wasn't really an organization at all. True, our group was doing excellent science, including contributing to the development of automated dating of single crystals, but we were living on so-called soft money, that is, funds from grants, with no long-term stability. We had no institutional structure providing benefits, no steady payroll. Because of our nonstatus, grants from, for instance, the National Science Foundation were officially awarded to IHO, for disbursement to the geochronologists. Our group was independent in our work and how we went about it, but we depended on IHO for our existence in the world of bureaucracy.

This loose organizational structure was made formal early in 1989, when the IHO board agreed to put the geochronologists on the institute's payroll, creating two arms to the institute: the anthropologists, with their director, and the geochronologists, with theirs. The anthropologists, with two staff scientists; and the geochronologists, with half a dozen. It was clear where the scientific vigor lay in this relationship. "We may have disparaged Don and Bill for not growing, for not expanding from their humble beginnings," says Paul Renne, who became director of geochronology in 1991, "but it was partly as a result of the administrative stability that IHO gave us that we were able to expand and grow. So we do owe them something for that." The price of stability, however, was the loss of a certain degree of scientific autonomy: from early 1989 onward, our group was to be known officially as the Institute of Human Origins Geochronology Center.

Tensions Build

When a crystal shatters, Freud said, the preexisting planes of weakness become clearly revealed. So it was with the breakup of the Institute of Human Origins. For instance, our group's view was that Johanson was spending far too much time on public relations activities—such as television appearances and writing popular books—to the detriment of fund-raising for the institute as a whole. At the same time, the anthropologists thought that we were spending too much time on science not relevant to the institute, such as the age of continental rocks and the timing of ancient mass extinctions. Each of these activities can be defended, of course. In the first case, raising public awareness of human origins research benefits the science as a whole, and the institute in particular. As for the second, scientific quests distinct from ones associated with anthropology are to be expected in a geochronology group of world-class distinction. Indeed, we had been told at the beginning of the relationship that these nonanthropological activities would be accepted as a natural part of what we would be doing. The fact that personal characterizations such as "He's just looking for public aggrandizement" and "You are all fucking nerds" were variously tossed back and forth at IHO therefore spoke as much to festering resentment as it did to the substance of the accusations themselves.

But there was more. In the five years leading up to the breakup, our group brought in more than 70 percent of the grant funds and authored more than 90 percent of the institute's scientific papers.[7] This could have been cause for celebration and pride among the other members of IHO. Instead, it seems to have been more a source of jealousy and efforts to belittle rather than recognize our achievements. And then, too, there is the question of personalities. Johanson and Garniss are about as different from each other as they could possibly be: the former, sartorial, overflowing with syrupy charm (when he chooses), and hungry for public recognition and acclaim; the latter, rumpled and grandfatherly in appearance and demeanor, and caring more for science than limelight.

And yet, as Clark Howell noted when he brokered the marriage between the anthropologists and the geochronologists, both men sport considerable egos, and, it has to be said, explosive tempers. In this second category, Johanson is clearly the more accomplished. In our years of collaboration at the institute he developed a well-deserved, if unfortunate, reputation for unpredictable and high-decibel personal verbal assaults, sometimes accompanied by the tossing around of books or whatever else came conveniently to hand. Often subsiding just as precipitously as they began, these outbursts came to be known around the institute as "Johanson's tirades." Garniss once asked Bill Kimbel why Johanson had picked so viciously on a certain individual, to which Kimbel replied simply: "Face it, Garniss, we all know Don is a madman. He's a *mad*man!" As things transpired, one such tirade would be a key trigger in the events that led to the breakup of the institute.

One final weak plane revealed in the broken crystal was a very personal animosity and jealousy that Johanson expressed toward Garniss. Although Johanson is the better-known scientific personality, Garniss is seen among scientists as the more accomplished scientist, and he enjoys the loyalty of his group. For instance, among other things, Garniss's contribution to the development of geochronology was recognized when he was elected to be chairman of the Eighth International Conference on Geochronology, ICOG-8, to be held at Berkeley in June 1994. Similar recognition in the realm of anthropology had eluded Johanson. "Don displayed his jealousy by trying to stop me making a presentation to the IHO board about ICOG,"

Garniss recalls. "And then when I did talk about it, Don would say things like, 'Garniss needs help with this. God, does he need help.' It was a rhetorical put-down."

Although Johanson from time to time took aim directly at Garniss in this way, more often than not he attacked others as a way of getting at Garniss. I was the target of one such incident when in 1990 I published a paper in *Science*,[8] a significant accomplishment for any scientist, but especially so for one in the early stages of his career. I made the mistake of giving my affiliation as the Berkeley Geochronology Center, which is what we were when I wrote the paper. By the time it came out we had changed to the Institute of Human Origins Geochronology Center, but I forgot to make the correction. Johanson hit the roof, saying that I was trying to undermine the institute. He insisted that I write to *Science* with a correction and to the IHO board members to apologize. Right about this time, most of the geochronology group went to Australia for a scientific conference. I stayed home; I was still trying to finish my dissertation. When everyone had gone, Johanson lit into me again, shouting and yelling at me, saying that I should resign on the spot. I said that it had been a mistake, not a deliberate slight, but he wouldn't listen.

When the geochronology group returned from Australia, bearing the news that Garniss would be chairman of ICOG-8, Johanson brought up my faux pas again, this time at a board meeting, saying that I should go. Garniss said that if Johanson insisted that I resign, the whole group would leave. "I think he was jealous that this young student was having a paper published in *Science*," Garniss suggests. "And I think he was jealous of the loyalty in our group. He longs for that."

But in no area was Johanson more jealous of Garniss than in Garniss's close relationship with Gordon Getty. Gordon has a long history of support for anthropology, through being a benefactor and chairman of the L. S. B. Leakey Foundation and the major financial backer of Johanson's institute. In 1989 Gordon pledged five years of support to IHO, in the form of a matching-funds grant to the tune of up to a million dollars a year. Johanson therefore depended heavily on Gordon's support, but it was Garniss who much more frequently was included in the Gettys' social events and trips, particularly in recent times. When Johanson confronted Gordon with the suggestion that he had funded our 1992 visit to Java, he was displaying his suspicion that Garniss had parlayed his relationship with Gordon into financial backing. And he was displaying how protective he felt about the institute's funding sources, wanting no one to interfere with them in any way. (As we explained in chapter 6, Johanson had forgotten that the modest funding for the trip—some $6,000—had been approved from the institute's own contingency fund two years earlier.) Johanson's naked display of jealousy served to nudge Gordon just a little closer to seeking changes in the way the institute was being run.

One such change occurred a year before the final break, when control of much of the institute's day-to-day operation was given over to a newly created Science Committee, made up of Clark Howell, Johanson, Bill Kimbel, Garniss, and Paul Renne, with Paul acting as chairman. The move to establish the committee was in part the result of Gordon's wish to reduce Johanson's overall influence on the institute. Johanson was then spending a lot of time—more than he had expected would be necessary—on a *Nova* television miniseries, "In Search of Human Origins," and was therefore less able than he should have been to shepherd the institute's affairs. But, in any case, Getty was coming more and more to the conviction that Johanson's volcanic personal style was incompatible with wise stewardship of any organization. The time Johanson spent on his television venture did become an issue in the rhetoric surrounding the breakup, but his brand of leadership was more pertinent to unfolding events.

The idea of a breakup of the institute cropped up more than once during its final few years, sometimes in the heat of an argument between the anthropologists and the geochronologists, sometimes among our group alone, over a few beers at the end of the day, when constant attrition of good faith culminated in outright frustration and a desire to be rid of entanglement. But then, in the fall of 1993, there was a collective effort to make a go of things, to try to put past antagonisms behind us. This was expressed publicly at a dinner after a board meeting in New York, in November. It was a "kiss and make up" kind of event, not unlike a final attempt to rescue a failing marriage. There was lots of embracing going on, lots of talk about what a good thing we all had going for us, so why not make it work. "I came back from New York with a new sense of desire to work things out," recalls Paul Renne, "agreeing with Bill that we were now on the right track."

But the spirit of optimism and mutuality didn't last long, and soon it existed as a facade only. We geochronologists sensed something conspiratorial developing in the other arm of the institute but didn't know what it was. Questions were being raised about the institute's finances—which had been through difficult times—and about Johanson's own finances in relationship to the institute, notably about his speaking fees and royalties from the book associated with the *Nova* television program. A confection of issues fed into a growing unease at the institute, including questions over access to research sites in Ethiopia and the awarding under unusual circumstances of grants in the institute's African Scholars Program. And there was continued tension over a member of our group who had joined as a Johanson loyalist but had never really fitted in, either socially or professionally.

Johanson continued to belittle our work, usually in private (that is, within the orbit of the institute), but also sometimes in public, including at a press conference held at the institute on the publication of Kimbel and Johanson's paper in *Nature,* on the 3-million-year-old *Australopithecus afarensis* skull, in March 1994. One of the reporters asked Johanson how he knew how old the skull was. Johanson's reply was to the effect that we used the technique of argon-40/argon-39 dating, which had been developed by Derek York in Toronto, and that our lab was a copy of York's. In fact, as Paul Renne was forced to repeat more than once, yes York contributed to the development of the technique we used; however, we ourselves played a big role in that development, and are widely credited for our work among our peers. For example, Alan Deino's software permitted the completion of the first fully automated argon-40/argon-39 dating system. Not so by Johanson.

Our patience with Johanson's continued slights had worn thin by this point, and Paul felt compelled to pen a memo to the institute's science committee, to put the record straight.[10] "At a recent press conference attending the publication of new fossils from Hadar, it was erroneously stated by Don Johanson that our dating methods were borrowed from the University of Toronto," Paul's memo began. "Several other misstatements have occurred within the last year or so. I want to clarify the relevant history in order to avoid any future such misstatements and preempt the need for correction." Then, in a clear expression of battle lines drawn, Paul went on: "This memo is also intended to serve notice that, henceforth, I will immediately correct any inaccurate statements about geochronology in the same forum in which they are made." No more sitting tight-lipped in public, listening to scientific misstatements and not correcting them for the sake of institute loyalty—that was Paul's message.

From the tone of this memo, and the depth of personal distrust and scientific antagonism that it reflected, it was clear that a rocky relationship could not go on much longer.

But it wasn't a dispute over science that brought matters to a head. It was a lunch.

Conflict Explodes

Chez Panisse, on Berkeley's Shattuck Street, is one of the most acclaimed restaurants in the United States. In the evenings, gourmets from around the country—and the globe—savor a daily-changing, no-choice dinner menu on the modestly appointed restaurant's ground floor, for which they pay high prices and must reserve weeks in advance. On the restaurant's second floor—Chez Panisse Café, as it is called—lunch of equally delicious fare can be had in more casual circumstances. It is a favorite spot for transacting discreet business conversations, a statement of class.

On April 12, 1994, a Tuesday, we had arranged to lunch at the Café, with Kay Woods and Barbara Newsome, board members of the Leakey Foundation, and Phyllis Wattis, a San Francisco philanthropist who had contributed generously to anthropology. A month earlier, I had talked about our work on the age of the Javan fossils to a group of Leakey Foundation people at Woods's house in San Francisco. Wattis, who had been unable to be there, wanted to hear about the work firsthand. An intimate lunch would work very well, she told Woods. Woods made a date for a lunch gathering at the Café.

We knew of Wattis's laudable record of philanthropy in science, including support for the Hall of Man at the California Academy of Science and contributions to the Leakey Foundation. What we did not know, however, was that a few days prior to the lunch, Johanson and Clark Howell had met with Wattis as a preliminary approach to seeking her financial support for IHO. The institute's executive director, Susan Shea, had put out a memo to the effect that Wattis was a potential donor for IHO, with a tacit reminder of institute policy: potential donors must not be approached by more than one representative of the institute at any one time. In other words, Keep Off!

Conversation at the lunch table, still in preliminary, casual mode, was interrupted after a few minutes when I noticed that Johanson, accompanied by his wife, Lenora, had stopped briefly at the head of the table. I said, "Hi, Don," and was about to introduce him to Phyllis, but he went right on by without saying a word, headed for the back of the Café at a fast clip. Then, toward the end of lunch, after we had described the Java venture to Wattis, the process was repeated, in reverse, but with Don pausing at the head of the table much longer this time, just standing there and glaring at each of us. Garniss offered a greeting this time: "Hi, Don." Nothing. No response. After a few seconds—which seemed much longer—Johanson exited, again at a fast clip. Stunned at first, the lunch party was soon laughing at Johanson's performance, speculating on what might have prompted it.

"We didn't know what it was all about," says Garniss now. "It's well known that Don doesn't care much for the Leakey Foundation, to put it mildly, so we thought maybe he was reacting to the presence of Kay and Barbara. I had no idea there was a problem with Phyllis being there." That supposition was wrong, as Garniss was soon to find out.

A meeting of the science committee was already scheduled for 2:00 that afternoon. But when we returned to the institute, we were summoned into Shea's office, before the meeting. I got there first, not knowing the reason for the invitation. Shea, obviously extremely agitated, mentioned lunch, saying we had gone behind Johanson's back. Shocked, I explained that we were there simply to talk about our work in Java. I suggested that she call them if she didn't believe me. When Garniss arrived, he was met with the same accusation, that he had ignored the memo about Wattis being a potential donor and that he was trying to undercut Don by asking for money separately. "I don't even remember seeing the memo," remembers Garniss, "and in any case it would have been irrelevant, because we weren't meeting to talk about money. We were at lunch to talk about science, our science." Garniss blew up at Shea, saying, "I couldn't care less if this upsets Don. We weren't there asking for money."

The scene repeated itself, only more explosively, when Garniss walked into the science committee meeting. Shea was there, as were Johanson and Kimbel. Paul was absent, as was Clark Howell. "Don jumped on me with all these accusations about my undercutting him and so on," Garniss recounts. "I repeated that we weren't seeking money, and that the meeting could actually help the institute in its goal, not hurt it."

"Garniss, where did you learn to lie so well?" Johanson shot back.

"If you don't believe me, go ask Kay or Barbara," Garniss responded. "They'll tell you what happened." But Johanson was in no mood to listen.

The encounter was as contentious as it could be, with Garniss's and Johanson's deeply rooted opinions of each other on full view, no window dressing, no attempt at soft-pedaling.

When, a few days later, Paul heard of the event, he called Gordon Getty and recounted what had happened, adding that there were a lot of problems at the institute and that "we are kidding ourselves over this supposed reconciliation." Gordon then phoned Garniss, and asked if it was true about Johanson's bizarre behavior in Chez Panisse. Yes, it was true, Garniss told him.

"I had respected Don as a scientist," Gordon now says, describing his reaction to the events. "I'd heard stories about his behavior, how he treated people, about how he has a tendency to blow up at people around him, except those he needed for money. He never blew up at me, for instance. For a long time I had put aside the stories I heard, the complaints, tried to ignore them." By itself, the Chez Panisse incident—as it came to be known—was not sufficient for draconian action on anyone's part. But, explains Gordon, because it came on top of many, many personal tirades and putative instances of mismanagement and unprofessional behavior on Johanson's part, he could no longer ignore what was going on. The Chez Panisse incident was emblematic. "Mean-mugging colleagues is pretty childish," says Gordon, describing Johanson's deliberate and unfriendly glaring at Garniss and me at the lunch table. "But mean-mugging Kay Woods and Barbara Newsome, two of the most eminent trustees of the Leakey Foundation—that shows he's not in control. This guy shouldn't be in charge of anything."

An Abrupt Parting of the Ways

On April 1, 1994, Tom Hill, chairman of IHO's board, had written to Gordon Getty about future funding for the institute. Hill acknowledged the importance of Gordon's five-year matching grant, which had started early in 1989 and was instrumental in bringing the geochronologists more formally under IHO's administrative wing, and which was due to expire in December. "The benefits of your support are coming to fruition," Hill noted. "Our scientists want the Institute to flourish; the Board wants that also." Appended was a long list of scientific publications during 1992 and 1993, of which 90 percent came from our arm of the institute, 10 percent from anthropology; and a list of field expeditions in the same period, split along the same lines as the publications. In a separate proposal, authored by Hill and Susan Shea, IHO's board sought a similar million-dollar-a-year matching grant from the Gettys, beginning in January 1995.

In Gordon Getty's thinking, the unfolding events triggered by the Chez Panisse incident exploded all plans for the next five years. Following his conversations with Paul and Garniss, Gordon undertook a fact-finding operation, calling people at the Leakey Foundation, at the institute, and on the IHO board. He decided that what he had discounted for so long could be ignored no longer—Johanson should step down as president, and perhaps even leave the institute altogether. Gordon resolved to withdraw his support completely if he was thwarted in this. As a clear warning shot of his seriousness, Gordon had Lawrence Chazen, director of the Ann and Gordon Getty Foundation, write a letter to Susan Shea, stating that, "effective immediately," the flow of matching funds would end. Gordon's intent was to ensure that the institute got back on what he considered the right track before he committed further funds; otherwise, the proper course of action would be, he thought, for the two arms of the institute to separate in an orderly fashion. An emergency meeting of the board was arranged to discuss the crisis, prior to which Gordon had several conversations with Tom Hill and David Koch with the object of exploring possibilities. Hill and Koch, possibly Johanson's greatest supporters on the board, and Gordon, by now Johanson's greatest detractor, had very different views of the past and of the way the future might unfold.

As the May 3 emergency board meeting approached, there was, remembers Paul Renne, "a sense of impending doom." The institute was tearing itself apart, and "we had no idea what would happen." We knew we had Gordon's goodwill, but there was nothing concrete, no promise of tangible support if events went badly. What would an orderly separation mean for us?

The meeting was called to order at 1:00 P.M. in the institute's library. Nine board members were present in person, four others were participating by telephone conference call, including Hill and Koch. A fourteenth member had arranged to vote by proxy. Hill opened the meeting by summarizing Gordon's 29 April letter that proposed his suspending matching funds, and asked if he had a proposal for consideration and a vote. Gordon responded by saying that he understood from earlier conversations that David Koch had a very specific proposal, to split the institute in two. Perhaps a committee could be formed to discuss that, Gordon suggested.

Asked why the emergency meeting had been called, Gordon explained that a year earlier Johanson had been demoted because of "tantrums" and said that there had been "another tantrum." Anyone else behaving like this would have been asked to leave after the first such incident, he said. Johanson should not receive special treatment. Now that Johanson had finished his *Nova* project and had no projects in hand in terms of grants, it would be an appropriate time for him to move on to other ventures, Gordon suggested. "I think he could leave with a minimum of embarrassment if all were of a mind to minimize embarrassment," he explained. "And that is my proposal." Asked if this was a condition for his future funding of the institute, Gordon responded, "Yes." Garniss seconded Gordon's motion.

Gordon was asked to clarify his position—how much support he would give and for how long. "We would have to work that out in good faith," he answered. "I would make no commitment at this time beyond carrying through to the end of the year, as originally envisaged." David Koch was not reassured. It looked to him as if the institute was going to run out of money, he said, so "we're going to have to take draconian measures, and cut costs to the bone." He proposed immediate "termination of the geochronology staff" as being necessary for "the survival of the institute."

Some board members expressed incredulity that the situation, though difficult, had come to this drastic point. Others protested that surely Johanson was not the only issue here, that in any case he founded the institute and no one around the table would be present but for Johanson's vision. Still others quickly got down to looking at the bottom-line numbers, how much money would be saved with our group out of the picture. In fairly jagged manner, discourse switched back and forth among these concerns. When he was invited to speak, Johanson said that the *Nova* venture had taken more time than he had expected, but that he was now anxious to get back to the science, back to the field. The institute's problems wouldn't vanish if Johanson were to leave, Bill Kimbel insisted. There were issues of philosophy and policy. "I cannot believe that we stand here on the precipice because of Don Johanson losing his temper," Kimbel ended. Tom Hill's crackly, disembodied voice on the speaker phone assured Kimbel that Johanson's behavior was not the only issue at hand. If Gordon Getty really believed that Johanson's removal would solve everything, said Hill, then there was no alternative but to follow David Koch's proposal and terminate the geochronology staff.

Gordon did not like the way the meeting was going, the sentiments expressed, and the way loyalties were lining up. He said he was prepared to work with a committee for interim support, until the matching grant ended in December, but that "there's no possibility of my making any commitment beyond that." Then, increasingly angered by the board's stance and the proposal to terminate the geochronologists, Gordon withdrew even his offer of support for the next few months. Any funds he gave to the institute—assuming that something could be worked out—would be solely for the geochronology group.

Through the fog that was descending on the path of the discussion, Gordon's proposal was described as "unreasonable." Koch's motion to eliminate our group immediately was formally put forward. Johanson seconded it. The board quickly moved to vote, first on Gordon's proposal, which was voted down, nine votes to four, with only Paul, Garniss, Clark Howell, and Gordon supporting it. On Koch's motion, the vote was again nine to four, this time in favor, with Paul, Garniss, Clark, and Gordon opposing it.

Susan Shea and the institute's attorney, James Carter, were asked to work out the practicalities of the termination, with the hope of avoiding litigation. Paul brought up the question of the fate of the equipment in the geochronology lab. "The equipment is owned by the institute," Tom Hill said bluntly. "If Gordon does want to set up an institution to pursue the science of geochronology, this committee would certainly entertain proposals." Gordon said to Garniss, cryptically, "I guess your guys are going to have to look for new jobs."

Needless to say, the presentation to the World Presidents' Council, planned for the day after the board meeting as a celebration of the institute's recent scientific successes, did not take place.

The Path to Court

The immediate aftermath of the board meeting was thick with anger, frustration, and disappointment. That evening Garniss and Paul drove across the Bay Bridge to Gordon Getty's house in San Francisco's Pacific Heights, where they joined Gordon, his wife, Ann, and a couple of Gordon's friends from the Leakey Foundation, including Bill Newsome. Gordon Getty was furious with IHO's board, and more than a little bitter that he had lost what effectively had been a game of political poker with David Koch and Tom Hill over the institute's future direction. Those who had been at the meeting went over events again and again, the way people do when there is a measure of disbelief over how circumstances transpired. Those who had not been present offered sympathy and validation. No real decisions were made, no specific plans formulated, although there was an implicit understanding of support by Gordon for the geochronologists' immediate future.

Meanwhile, Susan Shea and Bill Kimbel met at the house of IHO's attorney, with the aim of drawing up a joint-use agreement that would allow the geochronologists use of their facilities at the institute, on an interim basis, until June 30, and would require the geochronologists to find their own funding, presumably from Gordon. Shea invited Paul to join the meeting, but he didn't come because he needed to be at Gordon's gathering. Paul did talk to Shea on the phone a couple of times, however, and so gained a sense of the tone of the draft agreement. When, the following morning, he saw a completed copy, he said that our group must reject it "on advice of counsel," partly because it extended for just a month the group's access to its laboratories and thus offered no long-term security, but more pertinently because it would effectively have meant that we would forfeit any future legal action against the institute. And there was a lot of potential litigation to contemplate, not the least of it for wrongful dismissal.

Having had his proposals spurned so indecorously by Johanson's supporters on the institute's board, Gordon was in no mood for reconciliation or negotiation. As towering as he can be as a supporter, he is no less formidable as an adversary. "I consider the board's action unsavory and dishonorable, and resign from it effective now," he wrote to Shea the day after the board meeting. "IHO is welcome to negotiate with anyone who trusts them." It was clear that he, Gordon, did not. Gordon also indicated that he had "advised the fired employees to look for other jobs," but that he would support them, "from compassion [and for] as long as it takes."

Shea responded on May 5, graciously acknowledging the importance of Gordon Getty's previous support in the institute's success and urging him to reconsider his decision not to negotiate with the institute's board. "We don't stay angry forever," Gordon wrote back the same day. "I'm sure the board thought they were fighting on the side of the angels . . . But I'm afraid it's their mess to clean up. My decision not to negotiate with IHO, even through a trusted intermediary, is final. Life is too short." Gordon was taking the position that at the ill-fated board meeting he had essentially offered a joint-use agreement as a transition to an orderly separation, but that the board had rejected it and had chosen instead the drastic solution of instant dismissal of our group. Johanson's supporters on the board had apparently elected to vent their feelings about Gordon's withdrawal of funding for IHO by firing us on the spot, thus precipitating a crisis rather than solving one. "The board had brought this thing on themselves," he explains, "and it wasn't my place to help them out."

With Gordon's moral and financial backing, on May 6 our group incorporated ourselves as a nonprofit organization, the Berkeley Geochronology Center, thus reclaiming the name we gave up when we became part of IHO early in 1989. The speed with which our attorney, Cynthia Rowland, had prepared and filed the papers of incorporation was little short of remarkable, and displayed our group's determination to succeed on our own terms. Although we didn't have access to our labs and therefore could not continue our work, we were still permitted to use our offices, as well as a couple of extra rooms we managed to lease in the same building. This was important, because with ICOG-8 due to begin on June 6, Garniss, as the conference chair, needed to be able to complete final preparations, aided by an army of volunteers.

On the afternoon of May 10, the newly appointed board of BGC met with William Orrick, an attorney with the San Francisco law firm of Coblentz, Cahen, McCabe & Breyer, and asked him to seek a resolution of the outstanding legal issues with IHO and to work toward an acceptable joint-use agreement. During the following four weeks the exchanges between Orrick and Brian Carter, IHO's attorney, to these ends documented a steady widening of the gap that separated the parties, not a hoped-for narrowing. Mutual trust was scant between the principals—that is, IHO and BGC—and the language between the lawyers ever more caustic. "BGC . . . is a mere 'trespasser' [in IHO's facilities]," Carter wrote to Orrick on May 27. And "IHO, after almost 4 weeks of fruitless negotiations, will have no alternative but to declare negotiations to be at an impasse." Orrick responded to Carter by letter the same day, saying that "your client has not shown a shred of good faith from the time it summarily terminated the current BGC's employees without good cause." And "I assume from your letter that you are not rattling your saber, but that war is at hand." And on and on, with a mutually acceptable joint-use agreement remaining elusive.

The most incendiary issue came on May 23, just as Orrick and Carter felt that they might actually be making progress. On that day Susan Shea faxed a new draft agreement to Orrick, which included—out of the blue—the demand that BGC pay IHO $555,535 "for non-grant-funds and leasehold improvements." Gordon was outraged by the proposal, not least because some of the equipment in question he himself had bought, either as a special purchase or through his funding of the institute. "They were holding the scientists to ransom, holding the science to ransom," he recalls. "If it wasn't spite, it was extortion." Gordon 's temper did not improve when, later, IHO indicated that if he wouldn't ante up all the money, or an amount close to it, then the equipment would be sold, for scrap if necessary.

The crisis had reached an exquisitely tense point with each side enjoying strengths and weaknesses. IHO had access to all its facilities, but it was in a financial hole, and now faced the added expense of having to pay rent for space previously paid for by the geochronologists. BGC was well funded: Gordon had come up with a million dollars to establish the new entity and had promised to continue that level of support annually, and BGC researchers had recently been awarded more than $700,000 over three years in National Science Foundation grants. But we were denied access to our technical facilities. If we were to set up a new center elsewhere in Berkeley, that would be an expensive proposition; at least a year would be consumed by our having to prepare, test, and calibrate new equipment. It would be quicker and cheaper, obviously, if we were to buy our existing machines from IHO, as proposed, but Gordon adamantly refused to be pulled into a transaction he considered unseemly. Although we sympathized with Gordon 's reluctance, we were anxious to get back to work. Paul Renne met with Bill Kimbel on several occasions to explore what lesser figure might be acceptable to IHO, and then checked back with Gordon. Even at a quarter of a million dollars, Getty refused to budge.

Negotiations over a joint-use agreement continued into early June, with the lawyers again making some progress. Then the whole process was derailed again, this time by an explosion over ICOG-8. IHO had provided some logistical support for what was to be a big event for Garniss, and the institute had hoped for public visibility in return, and probably some small economic reward, too. On June 2 Garniss received a memo signed jointly by Shea, Johanson, and Kimbel, claiming that the programs for the conference, which had just arrived at the institute, made no mention of IHO. "The omission of IHO in the program is absolutely outrageous conduct," the memo hissed. "Our role as co-host, co-sponsor, financial and multi-year logistical supporter of the conference has been deliberately slighted . . . The ICOG-8 office within IHO space will no longer be available; neither will the library be available. You have abused your position as ICOG-8 chairman and have forfeited the claim we granted you to share our space." Locks were put on the doors the following afternoon by a giggling Johanson, who apparently savored it as a sweet moment of retribution.

"The conference was due to begin in just a few days, with nine hundred people coming into town," remembers Garniss. "Don knew locking us out of our offices was going to cause tremendous disruption. Plus, an important part of the event for many people was to have been the opportunity to visit the labs. Don was always jealous of our holding the conference, because he longed for that same kind of recognition. He just wanted to make it fail." Johanson even threatened to get an injunction to prevent the conference from opening, and said that he would have police cordon off the auditorium so that participants couldn't get in.

Some measure of negative reaction (although perhaps less dramatic) from Johanson over the omission of IHO's name from the ICOG program material would have been justified—if the allegation had been true. In fact, on at least seven items—including the program itself, the original invitation, postcards, registration material, and a questionnaire—the institute's name was prominently present. "I don't know what Don was thinking," says Garniss. "Was he deliberately not seeing it, just to make trouble? You tell me."

Whatever the motivation, trouble there certainly was. And it wasn't just with the conference. It was also in court.

A Battle Lost

On Wednesday, June 8, 1994, Case No. 736234-9 was heard in the Superior Court of the State of California, in the County of Alameda: to wit, "Complaint for breach of charitable trust, declaratory relief, accounting, and preliminary injunction." Plaintiffs, the Berkeley Geochronology Center; defendants, the Institute of Human Origins.

The primary goal of the action was to get us back to work in our lab at IHO, at least until the end of November, by which time we thought new facilities could be established. But the litigation's scope was much wider, not least because Gordon Getty was "fed up with the mismanagement of finances and personnel by the President and chief executive officer of IHO, defendant Donald Johanson."[11] Among other things in this respect, the complaint alleged that Johanson was retaining monies from book royalties and activities such as speaking engagements that, under IHO's articles of incorporation, should have gone to the institute.

The complaint stated the facts, such as the summary firing of our group, our being locked out of our lab, thus disrupting our work and ICOG-8, and the threat to sell the equipment. It also imputed motives. For instance, the complaint stated that by firing the geochronologists and threatening to sell the equipment, IHO was revealing that "its sole purpose [was] to injure BGC and its scientists."[12] The complaint also suggested that it was "not coincidental that IHO chose Friday afternoon before the international conference to lock BGC from the laboratory. It is doing the maximum to harm BGC and its scientists."[13] IHO's legal response was that "all of the potential harm could have been avoided had plaintiffs acted reasonably in accepting a proposed agreement offered by defendants immediately following the decision by the Board of Directors on May 3rd, when the employees of the Geochronology Division of the IHO were terminated under a Board resolution."[14]

The legal umbrella of BGC's complaint, however, had to do with the status of the institute—that it was a charity with responsibilities to the public trust. "The IHO raised funds for geochronology research, which constitutes a charitable trust," explains Orrick. There was close to $800,000 in current assets assigned to geochronology activities at the institute. "When IHO fired the geochronologists, discontinued geochronology research, and threatened to sell the geochronology equipment, that was a clear breach of charitable trust. Or so it seemed to us." Unfortunately for us, the judge didn't see it the same way. The temporary restraining order we had sought was not granted, and instead a date for arguing a preliminary injunction was set for mid-July. IHO did not allow us back into our lab to continue our work. But ICOG participants were allowed to visit the lab, guided by members of our group, two days before the conference ended.

Timing Is Everything

The decision in the June 8 court case had undoubtedly been a setback for BGC's immediate interests and long-term future, but Orrick refused to characterize it as a defeat. "Charitable trust cases are extraordinarily difficult to win, unless you get the support of the attorney general," he explains. "I had put our case to the AG, but had obviously not persuaded him of its merit. Our next step therefore was to build a better, more convincing case." This he attempted to do by, among other things, bringing in testimony from outside scientists, who attested to the damage done to science by IHO's actions. "We made our case, presented it to the AG. And waited," recalls Orrick.

Meanwhile, the issue of the fate of the geochronology equipment remained a constant irritant—and a possible source of resolution. "Even though the thought of our buying the equipment from IHO was morally repugnant and irritating beyond words, we began to see it as the most practical path," recalls Paul. "We were looking at potential space in Berkeley where we could establish a new lab, and we could see that the whole process would be much less expensive, and quicker, if we could have our own equipment without having to go through lengthy and costly litigation." Paul and Bill Kimbel had several phone conversations about the value of the equipment. Even though IHO wanted half a million dollars, the worth of the equipment on the open market was much less than that. If IHO was to be taken seriously by our side, they would have to trim their demands. And we would have to persuade Getty that, despite his pride, paying IHO a reasonable amount would be the most expedient and pragmatic way of resolving this issue.

As July wore on, the practicality argument finally began to win Getty's reluctant approval. Paul calculated that a fair figure would be around $120,000, and on the evening of July 14 Gordon agreed to offer this as the basis of a deal. Paul was to communicate the offer to IHO the following day. "I absolutely hated the idea," Gordon recalls, "but I could see the sense in it, so I voted Okay." That night Gordon tossed and turned, unable to sleep at all. "I just felt something was wrong, something didn't compute. That something was the attorney general. He had yet to make his decision. 'Why don't we wait for the next card?' I kept asking myself. I said to myself, 'Sure, we risk settling with IHO on worse terms than we were proposing, but then again, it might be much better.' That did it for me."

At 7:00 the following morning, the fifteenth, Gordon phoned Paul and Bill Orrick to tell them he had changed his mind, that he was prepared to wait for the attorney general's decision, whatever the cost. "They patiently and gently tried to reason with me," says Gordon, "and just as patiently and gently I held my ground." For Paul, Gordon's change of mind was depressing, because in it he saw the already slim chance of getting back to work slide away altogether. For Orrick it was back to business as it had been. What none of them knew, however, was that the attorney general had already made his decision.

On the afternoon of the day that Gordon had told Paul and Orrick of his change of mind, he went to the city for a series of appointments, as usual. "When I came back Meg [Starr] was jumping up and down with excitement," Gordon recalls with a big grin. "The AG had come in on our side like a ton of bricks." The arguments that Orrick had made had apparently convinced the attorney general that there were sufficient grounds for a case of breach of charitable trust by IHO, and he had filed a complaint of his own: *People of the State of California v. The Institute of Human Origins.* Soliciting public funds for geochronology research, and then summarily firing the geochronologists, subsequently failing to carry out geochronology research and withholding geochronology assets "constitute an abandonment of . . . charitable trust," stated the attorney general's complaint. The course of action he sought was "for a temporary restraining order, and preliminary injunction, transferring all geochronology funds and assets to an interim successor trustee and enjoining IHO from interfering with the charitable use of said funds and assets by the interim successor trustee."

"The timing of the AG's intervention was incredible," exults Gordon. "And he couldn't have stated more clearly that we were in the right." Gordon's sleepless night seemed serendipitous. "Gordon seemed to have had a hunch," says Paul, "and he was right, big time. The AG made himself a coplaintiff with us. How can you do better than that?" The attorney general also said that his office would investigate some of the financial mismanagement alleged in BGC's initial complaint. The attorney general's intervention effectively meant that our case against IHO was substantively over, even though it actually dragged on for almost a year before it was finally settled, out of court. Prior to that settlement, however, the fortunes of BGC and IHO changed rapidly and dramatically.

At 9:00 on the morning of July 22, our lawyers filed a motion in Superior Court for preliminary injunction, seeking to gain access to the geochronology lab space and offices and prevent IHO from disposing of any equipment. Three days later, Judge James Lambden signed the order and, moreover, stated that "Defendant IHO is hereby removed as trustee of said laboratory, equipment and offices, and BGC is hereby appointed successor trustee during the pendency of this action." Not only had we regained what we had previously enjoyed in terms of space and facilities, but also we, not IHO, were now the trustees. Orrick describes the result as "a monumental victory," not least because it had been a rare kind of case. "We were asserting ownership rights as trustees to equipment and leasehold space that was indisputably purchased by IHO, in the title of IHO, because we believed that they had broken the trust of the people of California and with their donors," explains Orrick. "There aren't many cases like this."

Garniss and I were enormously relieved when it all happened as it did, because we had been planning to go to Java in early August. "If the judgment had gone the other way, that would have been it for us," says Garniss. "As it was, the outcome and the timing were perfect. Just perfect."

Not so perfect, however, for IHO. In July, the board decided that the institute would move out of the premises it had occupied for more than a decade, and the move was made at the end of December. Ironically, the institute finished up in a building that we had once eyed for ourselves, in an industrial area of Berkeley, definitely a comedown from Ridge Road with its ready proximity to campus. Gordon's prediction that, in part because of its board's actions, the institute would face difficult times raising money apparently came true. "I was at an art exhibit in the summer of 1996, I think it was, and I ran into Jeff Meyer," recalls Clark Howell. Meyer had become IHO's chairman of the board. "He said the institute was having a tough time paying its bills and was likely to move." A year later, in the summer of 1997, the institute became a research unit of the College of Liberal Arts and Sciences at Arizona State University in Tempe, an arrangement that helped ease the financial burden, even if it did not constitute the establishment of a major center for the study of human origins, as Johanson had once dreamed.

Our lawsuit against IHO was settled out of court in May 1995, giving us "essentially all that we had initially asked for," as Paul puts it. The institute got relief from litigation, described by Susan Shea as "a moral victory."[15] Gordon had wanted to sue the institute for damages, but Orrick persuaded him that the attorney general would probably not look favorably on one charity suing another in that way. Gordon backed off. In a separate settlement, the attorney general ordered the IHO board to comply with proper accounting procedures in dealing with the institute's finances. He also directed Johanson and his wife to pay $24,221 to the institute, "representing the amount of revenue sharing that would have been payable had the IHO revenue sharing policy been fully implemented and enforced during the period January 1, 1990–December 31, 1994."

Meanwhile, not only were we reoccupying the space that had been ours before, but we were now using the anthropologists' labs and offices as well. A lot of energy had been expended and experimental time lost. But the divorce was now complete, and our Java ventures were back on track.

8

On the Cusp of Humanity

HOMO ERECTUS, the species to which the Mojokerto child belonged, stood on the cusp of humanity, no longer apelike nor yet fully human. In the evolutionary journey from the origin of the first member of the human family, some 5 million years ago, to the birth of fully modern humans in relatively recent times, the entrance of *Homo erectus* onto the scene marked a dramatic step in human prehistory. Every human species that preceded *Homo erectus*—that is, every species prior to about 2 million years ago—was distinctly like an ape in many respects: in anatomy, in important aspects of their pattern of growth and maturity, and in behavior. And every human species that followed *erectus* was distinctly human: in their anatomy, in their social lives, in their economic lives (subsistence), and in their mental lives. It was as if, metaphorically, *Homo erectus* carried a banner that proclaimed, "This way lies humanity!"

The number of "firsts" that *Homo erectus* can claim speaks volumes to the notion that a new kind of creature had arrived under the sun. *Homo erectus* was the first human species with a impossible to stand before the vibrant, chromatic images of the painted caves of Europe and not imagine you are looking at some expression of a distant people's perspective of themselves and their world.

In the modern, scientific world view, evolution has for many people replaced the creation myth as the explicator of who we are and how we came to be this way. In theory, by weighing the appropriate evidence—fossilized bones, stone artifacts, and, of late, genetic information encoded in our genes—we should come to an objective description of the course of our evolutionary history. In practice, however, we humans seem to find it very hard to be objective about a topic so dear to our hearts—that is, ourselves, and our place in the world of nature. We judge *Homo sapiens* to be something special in the world, and in many ways we are; and we infer that, because we judge our species as being special, the course, and causes, of our evolutionary transformation must also have been special in some way, embodying values that society holds to be important. You don't have to look very far back in the literature to find professional anthropologists expressing these positions boldly and passionately.

For instance, the British anthropologist Sir Arthur Keith wrote the following, just half a century ago: "Why, then, has evolutionary fate treated ape and man so differently? The one has been left in the obscurity of its native jungle, while the other has been given a glorious exodus leading to the domination of earth, sea, and sky."[1] One species skulks in "obscurity," the other exults in "dominion." Keith's contemporary Sir Grafton Elliot Smith had a ready explanation for these different fates: "[Our ancestors] were impelled to issue forth from their forests, and seek new sources of food and new surroundings on hill and plain, where they could obtain the sustenance they needed." The apes' inferior fate was in their own hands, opined Elliot Smith. "The other group, perhaps because they happened to be more favorably situated or attuned to their surroundings, living in a land of plenty, which encouraged indolence in habit and stagnation of effort and growth, were free from this glorious unrest, and remained apes, continuing to lead very much the same kind of life (as Gorillas and Chimpanzees) as their ancestors since the Miocene or even earlier times." To emphasize the point, he added that "while Man was evolved amidst the strife with adverse conditions, the ancestors of the Gorilla and Chimpanzee gave up the struggle for mental supremacy because they were satisfied with their circumstances."[2]

In Keith's and Elliot Smith's world view, evolutionary progress—and it is viewed as progress, not merely change—is the outcome of our ancestors doggedly pressing through adversity, the result of the good and valued Puritan ethic that nothing worthwhile is obtained without effort and struggle. The awful fruits of apathy, of being satisfied with what you have, are obvious. In other words, notes the Duke University anthropologist Matt Cartmill, putting such sentiments in historical context: "Darwinian man is lord of the earth, not because of any God-given stewardship or romantic affinity to the World Spirit, but for the same good and legitimate reason that the British were rulers of Africa and India."[3] While all this might sound amusing to our modern minds and our prevailing accepted wisdom, it is worth remembering that when Keith and Elliot Smith were writing their words, they were expressing what modern minds of their day considered to be accepted wisdom, too. Although anthropologists today no longer indulge in such putatively objective but in fact value-laden science, our thoughts about human prehistory are probably colored by social issues, as well as by science, in ways that we are not yet aware of but that will be blindingly obvious to scholars fifty years hence.

One thread of sociological influence that has persisted from Keith's and Elliot Smith's time on through to modern writing on human origins—albeit a thinner thread than it once was—is the notion of evolution as a journey, of an ancestor setting out on an evolutionary quest to reach some defined end point. For Elliot Smith, the course of human evolution was "the wonderful story of Man's journeyings towards his ultimate goal" and "Man's ceaseless struggle to achieve his destiny."[4] Compare Elliot Smith's words, written in 1924, with those of two anthropologists at the State University of New York, Stony Brook, penned sixty years later on the issue of the mode of walking by the earliest known (at that time) human species, *Australopithecus afarensis*: "In our opinion *A. afarensis* is very close to a 'missing link.' It possesses a combination of traits entirely appropriate for an animal that had traveled well down the road toward full-time bipedality."[5] Less romantic or evocative than Elliot Smith's words, perhaps, but our ancestor *Australopithecus afarensis* is still viewed as being embarked on a journey, marching toward being a full-fledged upright walker, just like modern humans.

There are several reasons why the temptation to think of a species' evolutionary history as a journey is so tenacious. The first is that, in hindsight, the events that led from ancestor to modern descendant can be strung together as if each was inevitably meant to lead to the next, rather than their being simply a series of evolutionary stages through time. In truth, evolution acts blindly and in the moment, and each species in an evolutionary lineage—each step in the journey through time—is discrete and complete in itself, an end in its own right, not merely a stepping-stone to something better. Second, the ethical values that Keith, Elliot Smith, and others expressed so boldly—that nothing worthwhile is attained without a struggle—is so very deeply ingrained in our thinking that the notion of the hero's journey is hard to eschew. Lastly, we humans *love* stories. Storymaking and storytelling is part of our being, an evolutionary endowment that allowed us to make sense of our world. Even the most objective description of an evolutionary history can therefore fall prey to the temptation of storymaking, even if it is the history of flatworms—but *especially* if it is our own history.

This little excursion through some of the sociological byways of paleoanthropology is meant as an alert signal about the way our ideas and our language are shaped when we think about humanity's place in nature. See, for instance, the opening paragraphs to this chapter, where the word "journey" is used to describe our evolutionary history and humans are noted as being "special" in the world of nature. As long as we understand the metaphorical use of the journey image, as long as we understand that by special we really mean, simply, "different," then we are likely to delude ourselves less than we might otherwise about what we are as a species and how we came to be this way.

Our Closest Relative

Children of all ages love to try to engage the attention of chimpanzees at the zoo, making faces, laughing at them, trying to get them to smile or "talk." We even dress up chimps in clothes, and have them conduct demure tea parties. They seem so much like us in many ways, and of course they are, for the very good reason that they are our evolutionarily closest relatives. But just how close that relationship is—as has become clear just recently—would have caused anthropologists of not so very long ago to reach for some steadying remedy. In the early decades of the last century, while most anthropologists were prepared to admit that there was something apelike in our ancestry, they took pains to ensure that our simian connection was *extremely* distant indeed. For instance, Henry Fairfield Osborn, director of the American Museum of Natural History in the early decades of the twentieth century, wrote in a letter to Arthur Keith in 1927: "I am perfectly confident that when our Oligocene ancestor is found, it will not be an ape, but it will be surprisingly pro-human." In today's geological framework, Osborn's putative human ancestor would have lived some 30 million years ago, and still it was more human than ape, in his eyes. "It may be called pithecophobia, or the dread of apes—especially the dread of apes as ancestors or relatives," was how William King Gregory, a colleague of Osborn's at the museum, described Osborn's sentiment in 1927.[6]

Osborn's view—the majority expression of the time—was perhaps surprising, given that half a century earlier Charles Darwin embraced our close relationship with chimpanzees (and with the other African ape, the gorilla). In his 1871 volume *The Descent of Man,* Charles Darwin stated that the African apes (gorillas and chimpanzees) were evolutionarily closer to humans than were the Asian apes (orangutans and gibbons), based on a comparative study of these species' anatomy. It was on this basis, incidentally, that Darwin concluded that Africa was "the Cradle of Mankind," as he put it. His friend and champion Thomas Henry Huxley was of the same opinion, as he argued in his 1863 book *Man's Place in Nature.* Despite the weight of these conclusions, anthropologists during the first half of the twentieth century chose to classify apes and humans in the following way: chimpanzees, gorillas, and orangutans were placed in one taxonomic family, the Pongidae, while humans occupied the family Hominidae in splendid isolation. This classification implies that the African apes are evolutionarily closer to orangutans than to humans, which is not what Darwin and Huxley argued at all. One reason for this anti-evolutionary classification was that anthropologists judged the apes to be more similar to one another in their daily life than they were to humans in theirs.

It took the appearance of a new science—molecular anthropology—on the scene to bring anthropologists back to Darwin's view, and to accept that our evolutionary relationship with apes is very close indeed, and recent. The term molecular anthropology was coined in 1962, by Emile Zuckerkandl, a colleague of Linus Pauling's. The two men had been pioneering the technique of inferring the evolutionary relationship among species by examining differences in the structure of certain molecules (proteins, initially) in the species. The technique is based on the fact that when an ancestral species divides into two daughter species, the genes in the two now-distinct lineages will independently accumulate changes, or mutations, over time. The longer the time since the two species diverged, the greater the difference will be in the genes and the proteins for which they code between the two species. Measure the differences and you may be able to reconstruct the evolutionary history. Whereas conventional anthropologists seek clues to evolutionary relationships through comparing differences in the structure of fossilized bones of extinct species, molecular anthropologists do it by comparing molecules from living species. Simple in principle, molecular anthropology is hard—and very complex—in practice.

In the early 1960s, therefore, conventional wisdom in paleoanthropology classified chimpanzees, gorillas, and orangutans as each other's closest relatives, with humans out on a limb, and estimated the time of origin of humans as at least 15 million years ago, and probably closer to 30 million. Then molecular anthropology intervened. First Zuckerkandl, and then Morris Goodman, of Wayne State University, concluded from comparison of certain protein interactions that gorillas and chimpanzees are evolutionarily closer to humans than to orangutans, just as Darwin and Huxley had said. Then, in the late 1960s, Allan Wilson and Vincent Sarich, of the University of California, Berkeley, estimated the time of divergence between humans and apes at a mere 5 million years ago. Not surprisingly, these conclusions were not well received, to put it mildly, because they were so at odds with prevailing anthropological thinking. It took twenty years before this view of human phylogeny became the new conventional wisdom rather than scientific heresy.

But there were more surprises in store. When they finally came to accept the molecular biological version of human history, anthropologists assumed that chimpanzees and gorillas were each other's closest relative, with humans a little more distant. After all, chimpanzees and gorillas are much more alike anatomically than either is like humans. Nevertheless, the molecular data—in the form of protein interactions and, increasingly, DNA sequence data—were initially equivocal on the matter. Some observers believed this indicated that there had been a three-way evolutionary split between us and the African apes, or, if not, then too close to call. Then, in the mid-1980s, the steady flow of DNA sequence and other genetic data began to yield a new message: that chimpanzees are evolutionarily closer to humans than they are to the gorilla. This was particularly hard for anthropologists to accept, for two reasons. First, the overall morphological similarity of chimpanzees and gorillas, including their very special anatomical adaptation to so-called knuckle walking (with weight borne by arms on bent knuckles), seemed to suggest a close relationship. Second, if chimpanzees are indeed closer to humans than they are to gorillas, it would mean that one or the other of two evolutionarily unattractive explanations would have to be adduced. Either the common ancestor of humans and apes used knuckle walking, a behavior that was lost in humans and for which anatomists see little or no vestige in our lineage; or knuckle walking evolved independently in the chimpanzee and gorilla lineages, an evolutionary coincidence that anthropologists are loath to accept. Such vestigial evidence has recently been pointed to in the wrists of two early australopithecine species.

Loath to accept it though anthropologists may be, that's what they seem to be facing as molecular data pile up in favor of the unlikely relationship. At latest count, 80 percent of experimental results fall in the human-chimp column, while only 20 percent support a greater evolutionary intimacy between chimps and gorillas. Through gritted teeth, most anthropologists have now come to accept the notion that chimpanzees are indeed our closest evolutionary relatives, with 99.5 percent of our genes being identical. Gorillas are more distant cousins. "Genetically," says Morris Goodman, "humans are only slightly remodeled apes."[7] Goodman is also pushing anthropologists to go the next step beyond accepting evolutionary reality, to incorporate it into their classification of apes and humans. In a nutshell, classification (a large swamp of a topic) is biologists' way of describing relationships among species. For instance, two closely related species may be classified in the same genus, such as the two species of chimpanzees: common chimps, known as *Pan* (the genus name) *troglodytes* (the species name), and bonobos, or pygmy chimps, known as *Pan paniscus.* Gorillas, which are evolutionarily more distant, have a genus classification of their own, known through a striking flash of nomenclatural imagination as *Gorilla,* the full genus-and-species appellation being *Gorilla gorilla.* Although they have different genus names, chimps and gorillas are classified in the same family, known as the Pongidae, which, as mentioned above, is also occupied by orangutans.

Traditionally, humans have been in splendid classificatory isolation, the sole member of the genus *Homo* and the sole member of the family Hominidae. Although many anthropologists are willing to consider putting humans, chimpanzees, and gorillas in the same family, to the exclusion of orangutans, in recognition of the known genetic relationships, that is not far enough for Goodman. Given our extreme genetic intimacy, says Goodman, humans and chimpanzees should be not only in the same family but also in the same genus, namely *Homo.* The distinction would then be between humans, as *Homo (homo) sapiens,* and chimpanzees, as *Homo (pan) troglodytes* and *Homo (pan) paniscus.*[8] As we contemplate ourselves as a species, and our place in the world of nature, think how very different is this perspective (sharing a genus with chimps, and having a common ancestor close to 5 million years ago) compared with Osborn's (in different families, and separated by at least 30 million years).

People who suffer from pithecophobia would surely have a very hard time with the modern anthropological world view of our relationship with the rest of nature.

The Ape That Stood Up

The following account of human origins held sway for decades, among professional anthropologists and in the public mind. Once upon a time, a very long time ago, a species of unusual ape in Africa was forced out of its traditional forest home because a cooling climate was steadily reducing the forest cover. The resourceful ape transformed adversity and potential disaster into ecological and evolutionary opportunity: once in its new, open savannah habitat, the ape at once began to undergo a series of evolutionary changes that adapted it to its new circumstances in a remarkable way, an evolutionary path down which no other ape had ever ventured. Gradually our ape-ancestor came to stand and move on two legs, not four; to make and use stone tools and weapons with which to hunt prey; to reduce the size of its long, sharp canine teeth; and to enlarge the size of its brain.

Once this evolutionary transformation was under way, it began to accelerate, pushed ever faster by the very changes themselves that were taking place, a positive feedback loop: the more the man-ape stood upright, the more it could use its hands; the more it used its hands, the more it needed to be upright; the more intelligent it became, the more it could rely on stone-tool technology; and the more extensively it used technology, the more its intellect was honed. Eventually our ancestor became a primitive version of us: erect and intelligent, a skilled toolmaker, an accomplished hunter. It stood triumphant on the African savannah, with less resourceful apes eking out a simple existence in the receding forests.

The notion that evolutionary advance was to be forged only on the anvil of effort was an explicit part of this scenario, as mentioned earlier. A second notion was that the evolutionary transformation from ape to human was effectively instantaneous, because the three qualities we take as separating us from the apes—upright walking, tool making, and high intelligence—all began to appear right at the beginning. In other words, the first member of our family was already identifiably like us, albeit in primitive form.

It was Charles Darwin who established this idea of concerted evolution in our history, with upright walking, tool making, and increased intelligence becoming ever more developed in lock step. In the almost complete absence of fossil evidence, Darwin had erected a plausible outline of our prehistory, emphasizing key characteristics of humanity as the prime movers in the evolutionary transition from ape to human. "If it be an advantage to man to have his hands and arms free and to stand firmly on his feet, of which there can be no doubt from his pre-eminent success in the battle for life, then I can see no

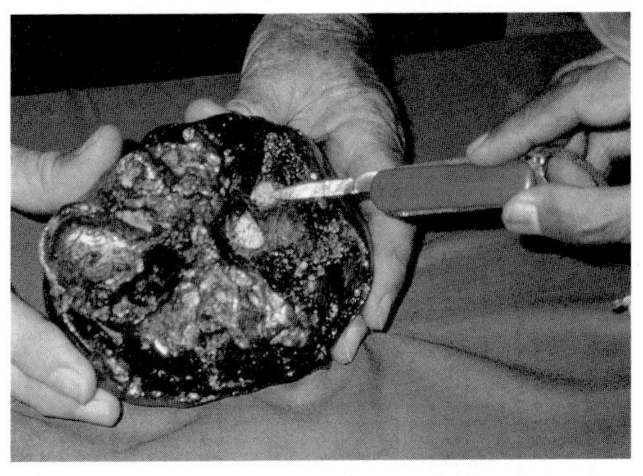

The tip of Garniss's penknife shows where Carl had scraped away some of the black surface of the underside of the skull. [S. Shute]

The most complete *Pithecanthropus* cranium found so far in Java, in the Sangiran region. (It is known technically as Sangiran 17.) [K. Barrett]

Sorting crystals in preparation for single crystal laser fusion dating. [D. Smeltzer]

Teuku Jacob points out the level at which he once believed Andojo indicated that he found the Mojokerto child's cranium. [C. Swisher]

The Mojokerto child's cranium. [C. Swisher]

The Trinil stone, on which is inscribed, "P.e. 175 m. ONO 1891/93." This cryptic message stands for, in Dutch, "*Pithecanthropus erectus wurde 175 meters Ost-nord-ost von dieser Stelle gefunden in den Jahren 1891/1892.*" Translated, it says, "*Pithecanthropus erectus* was found 175 meters east-north-east of this spot in the years 1891/1892." The stone stands outside a recently built museum on the river bank high above the excavation site. [S. Shute]

One of the Ngandong skulls (Skull VI) just prior to excavation on 13 July 1932. C. ter Haar, at right, went on to excavate the specimen.

There is a cottage industry in Java that produces fake fossils, usually made from bone. Here Carl holds up a supposed *Pithecanthropus* jaw, being sold at a gift shop near the museum in Sangiran. [R. Lewin]

Professor Raymond Dart examines the Taung skull at his office at the University of the Witwatersrand, Johannesburg, South Africa, in February 1925, not long after the skull had been discovered. Like the Mojokerto fossil, the Taung skull was also that of a child. This specimen was the first ancient human relic to be discovered, and it was given the scientific name *Australopithecus africanus*, southern ape from Africa. [Barlow/Rand]

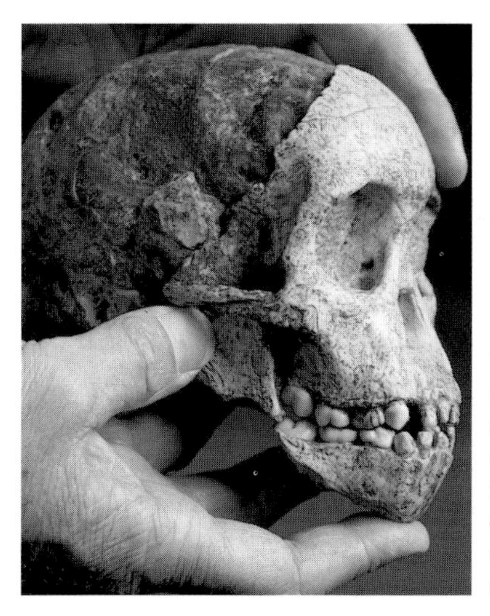

The Taung skull in more detail. Notice that the specimen is in two parts: the fossilized partial cranium, face, and lower jaw; and a natural endocast of the brain, that is, mineral that filled the skull soon after it was buried, hardened to solid rock, and recorded the brain's shape as impressed on the inner surface of the skull. [P. Kain/Sharma]

This adult specimen of *Australopithecus africanus* (Sts 5) was found at the Sterkfontein Cave in South Africa in the late 1930s, by Robert Broom. [P. Kain/Sharma]

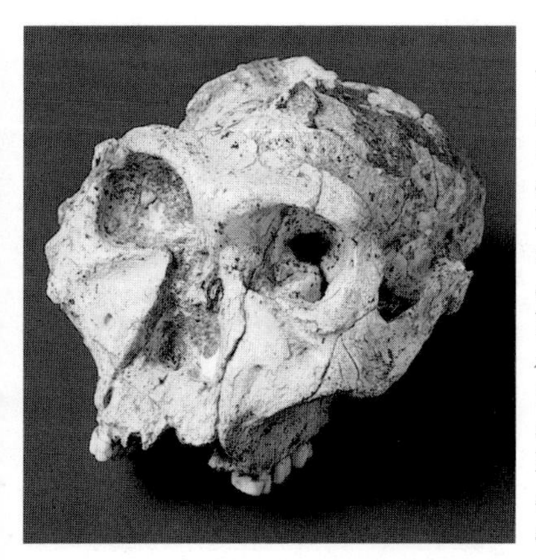

When specimens such as this one were found at the Swartkrans Cave, South Africa, in the 1940s, it showed that several different kinds of early human species had coexisted some 2 million or so years ago. This specimen represents the species known as *Australopithecus robustus,* so-called principally because its jaws were much bigger, and sported larger cheek teeth, than in *Australopithecus africanus,* which indicates that the two species lived on different diets; it has been suggested, for instance, that *africanus* may have had a more omnivorous diet than *robustus,* although certain analyses suggest that both species ate meat at times. [P. Kain/Sharma]

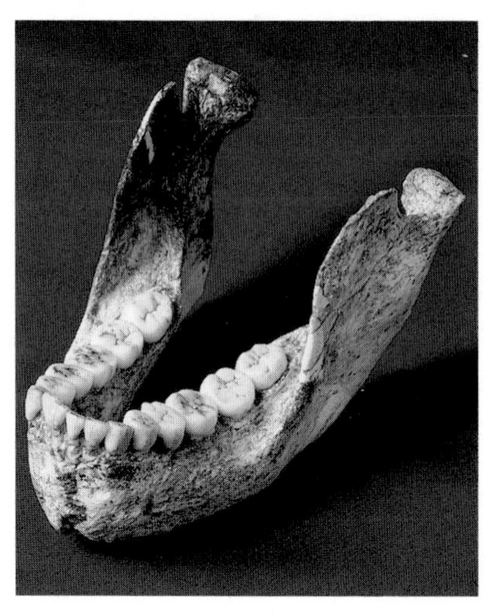

This jaw of an *Australopithecus robustus* shows just how big the cheek teeth were. [P. Kain]

When this skull (known as KNM 406) of an obviously robust species of *Australopithecus* was found in northern Kenya in 1969, it was further demonstration that early human species were diverse. Although robust in structure, like *Australopithecus robustus,* this Kenyan individual was sufficiently anatomically different from its South African counterparts that it was named a different species, *Australopithecus boisei.* [P. Kain/Sharma]

This skull, known as KNM 1470, was the first specimen of a large-brain, early human species to be found, in 1972, in northern Kenya. It lived at the same time as the robust and less robust *Australopithecus* species in the region, further emphasizing the diversity of human species that coexisted around 2 million years ago. [P. Kain/Sharma]

Specimen KNM 3733, found in Northern Kenya in 1975, lived almost 2 million years ago and was among the population that expanded out of Africa and into Asia. Some anthropologists call this type of African species *Homo erectus,* like similar specimens in Asia, including in Java; but others suggest it was a separate species, known as *Homo ergaster.* [P. Kain/Sharma]

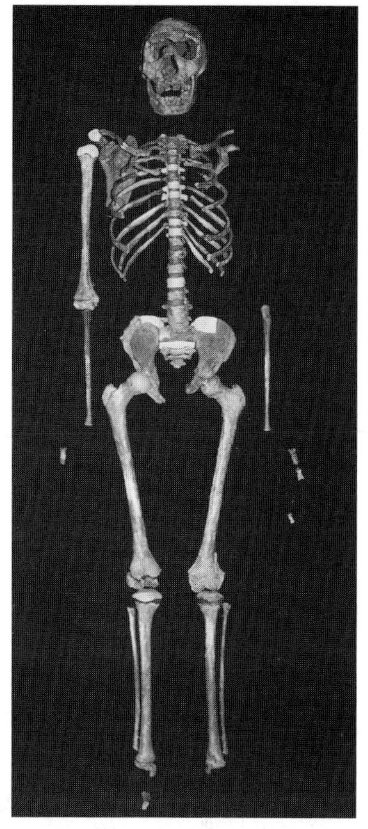

Known as the Turkana boy, this virtually complete skeleton of *Homo ergaster/erectus* was uncovered in northern Kenya in 1984. It reveals that this species had a long, rangy, humanlike body that would have equipped it for long-distance travel, such as the steady migration beyond Africa. [A. Walker]

Ancient technologies: The top two rows represent the kinds of simple stone implements that first appear in the archeological record almost 2.5 million years ago, and continue to be exclusively the prehistoric stone age kit until about 1.5 million years ago. At this point, larger, more complex tools appear, such as hand axes and picks, shown in the lower part of the picture. The simplest tool kit is known as the Oldowan technology and the more advanced as the Acheulean.

When modern humans evolved, a hallmark of their new skills was the stone tool technology, which included fine blades, as shown here. [R. Lewin]

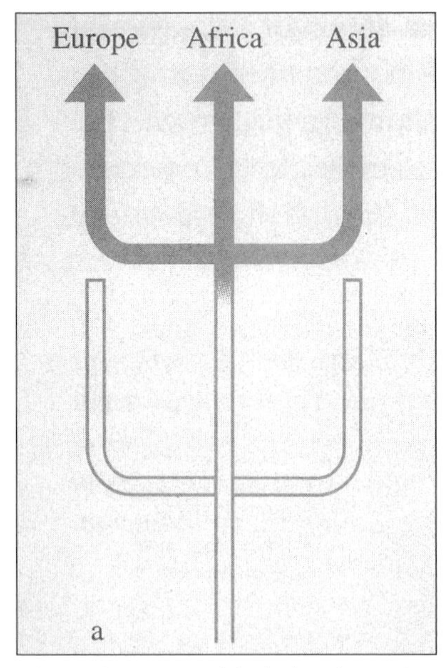

The single origin model, which indicates that many geographically local populations in Europe and Asia were replaced by recently evolved populations of modern humans that spread out of Africa.

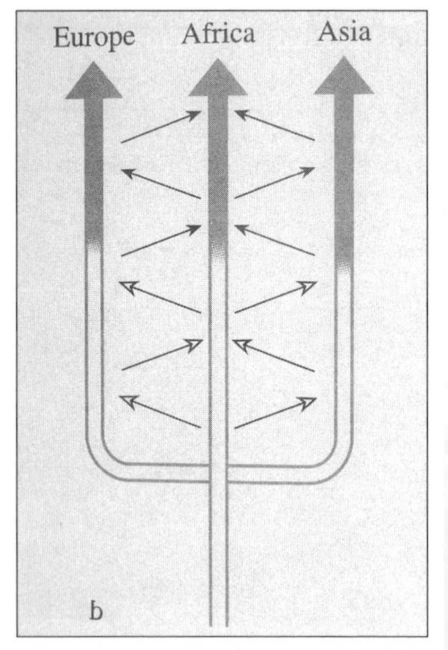

The multiregional evolution model of the origin of modern humans, which suggests that modern human populations evolved from ancient, geographically local populations, with a lot of interbreeding among them (indicated by arrows).

reason why it should not have been more advantageous to the progenitors of man to have become more and more erect or bipedal," Darwin wrote in 1871, in his *Descent of Man.*[9] "The hands and arms could hardly have become perfect enough to have manufactured weapons, or to have hurled stones and spears with true aim, as long as they were habitually used for supporting the whole weight of the body . . . or so long as they were especially fitted for climbing trees."[10] The small, bipedal, weapon-wielding, savannah-hunting human ancestor was now in a position to develop greater intelligence through more intense social interaction, said Darwin. And the large canine teeth would disappear too. "The early male forefathers of man were . . . probably furnished with great canine teeth; but as they gradually acquired the habit of using stones, clubs, or other weapons, for fighting with their enemies, or rivals, they would use their jaws and teeth less and less. In this case the jaws, together with the teeth, would become reduced in size."[11]

Remember that Darwin assembled this scenario simply knowing what modern humans look like, making assumptions about what the last common ancestor between humans and apes might have looked like (some kind of generalized ape), and imagining how the transformation from the one to the other might have occurred. He had no fossil evidence to guide him, no fancy techniques to aid him—simply his fertile and innovative imagination. It is therefore not surprising that he got some things wrong. But he got the essence right, and he clearly hit on an evolutionary package that so strongly resonated with anthropologists' thinking that it persisted as the dominant theoretical model for almost a century. That package—the simultaneous evolutionary advancement of bipedal locomotion, intelligence, and technology—meant that human ancestors were identifiably human in all respects, right from the start. It was, for instance, part of why Henry Fairfield Osborn could aver that our ancestors of some 30 million years ago were already more like us than like apes. It was a statement that because humans are special, the course of human evolution must have been special, too.

You have to go back only a little more than two decades to see this same sentiment—implicitly contained, not explicitly stated—in a prominent theory of our origins. Known as the Single Species Hypothesis, this evolutionary scenario argued that at any point in human prehistory there existed only one species. The course of our evolution, argued Loring Brace and Milford Wolpoff, anthropologists at the University of Michigan and the theory's principal proponents, was the sequential transformation of a series of ever more advanced species, beginning with an apelike progenitor and ending with modern humans. At the time—in the late 1960s and early 1970s—early human fossils had been unearthed in South Africa and in Kenya and Tanzania. Some of these specimens had ape-sized brains, others had somewhat expanded crania; some had small cheek teeth, others had enormous, millstone-like molars; some were impressively robustly built in their anatomy, while others were relatively slightly built.

Given this great range of anatomical variation, how were Brace and Wolpoff able to argue the Single Species Hypothesis? First, because of uncertainties about the ages of some of the fossils, this was not always possible to say which specimens were exact contemporaries, and in any case they were spread over a large area of a large continent. And we know from the stories in this book how big an impact uncertainty about fossil dating can have on evolutionary scenarios. The second reason, however, was more revealing. Even when specimens of very different anatomical features were accepted as being contemporaries, it was explained as anatomical variation *within* species, not variation *between* species. No matter that the degree of variation encountered among these petrified human relics was far greater than is typically encountered for within-species variation in the rest of nature. No matter, too, that the typical pattern of evolutionary change through time is simply not ladderlike, that is, progression from one species to the next, as the Single Species Hypothesis suggested, but instead is distinctly bushy, starting with a single, founding species, descendants of which split repeatedly, producing many evolutionary side branches in the newly emerging group. It is as if, in the rest of the world of nature, evolution, having hit upon some kind of novelty, quickly experiments with many variants on the new theme, producing a very bushy-looking family tree. Biologists call this an adaptive radiation. Bushes, not ladders, are nature's way of exploring novel possibilities in life.

Why, then, should humans be different—special—in this respect? Simply because we *are* special, was the answer given by, or implied by, proponents of the Single Species Hypothesis. Culture was one thing that made us special, they argued, and this changed the evolutionary dynamic, changed the evolutionary rules. It was a great equalizer, they said, so it was impossible to have two culture-making, protohuman species coexist. Unfortunately for the hypothesis's proponents, two major impediments were to render their brainchild moribund. First, the theoretical argument, borrowed from ecology, that two similar species could not coexist turned out to be overstated, at best. Second, the discovery in Kenya in 1975 of a large-brained, small-toothed individual (a type of *Homo erectus*) that was unequivocally contemporaneous with a small-brained, massive-toothed individual (of the species *Australopithecus boisei*) pushed the arguments of the hypothesis beyond credibility. The two creatures—who had lived not far from each other on the eastern shore of Lake Turkana, in northern Kenya, about 1.8 million years ago—were simply *so* different anatomically from each other that even the most ardent supporter of the hypothesis found it hard to maintain the argument of *within*-species variation. This, without question, was *between*-species difference.

The human evolutionary tree, as described by anthropologists, got bushier with that discovery. And with every passing year it has gotten bushier still. When Donald Johanson and Tim White redrew our family tree after they announced their new species, *Australopithecus afarensis,* it sprouted a few more branches, but it was a pretty spindly specimen, compared with the pattern of most species' groups in nature. And, having been blessed with what amounts to a flurry of discoveries of new early-human species in the past several years—in Kenya, Ethiopia, and Chad—the early part of the human family tree is beginning to exhibit respectable bushiness, in biological terms, that is.

There is still a lot of discussion as to how these early species relate—in evolutionary terms—to one another; more finds of these same creatures, and possibly of more new species, will help resolve that issue (see diagram, page 169). But in the context of the current discussion, several important points become clear. First, the path of human evolution—in its early stages, at any rate—conformed to the pattern most typical in nature: that is, an adaptive radiation of species, with nature experimenting on a novel theme. As a human family, we may judge ourselves to be blessed with certain special characteristics, such as language and consciousness, but we started out simply as an interesting evolutionary experiment. It was an experiment that ultimately led down some unusual evolutionary byways, but there was no guarantee that it would, only the possibility that it might, if prevailing ecological circumstances provided the right opportunity at the right time.

The evolutionary novelty that was the basis of the adaptive radiation that eventually gave rise to *Homo sapiens* was an

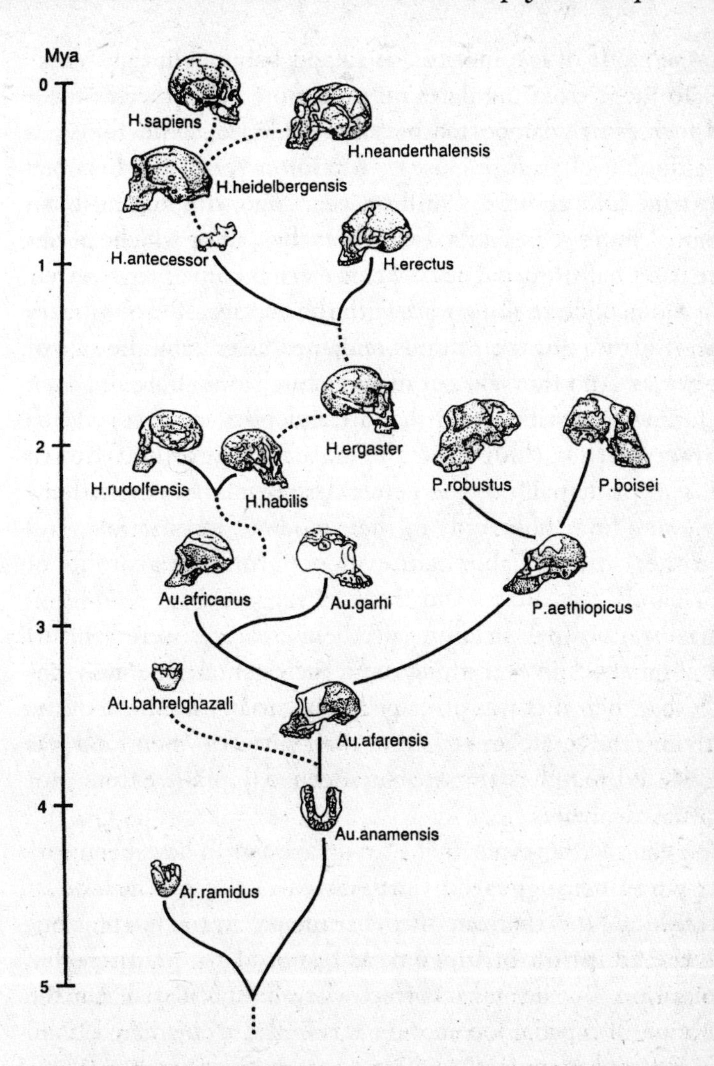

One interpretation of the human family tree based on currently known human species. The species themselves are unimportant here. More important is the sense of a bushy evolutionary record of human prehistory, rather than a steady, inevitable progression toward modern humans. With more fossil finds, this picture is likely to become more complex yet. (Courtesy of Ian Tattersall)

unusual mode of locomotion for an ape: habitual upright walking. To judge from the dates inferred from genetic evidence—and increasingly supported by fossil evidence, albeit flimsy as yet—this novelty arose close to 6 million years ago. Between that time and about 2.5 million years ago, the human bush sprouted more than half a dozen branches, all of which species were essentially bipedal apes. Their teeth became progressively more humanlike in some ways with the passage of evolutionary time, it's true. But their brains remained essentially the size of ape brains. And they did not manufacture stone tools, not even of the most primitive kind that archeologists can recognize as artifacts rather than sports of nature. These half-dozen species—principally of the genus *Australopithecus,* or southern ape—were humanlike only in their mode of locomotion. And even then, there is abundant evidence from the anatomy of their hands, feet, limbs, and trunk to suggest that, in addition to having a bipedal, striding gait, these creatures were habitual tree climbers. And everything about their dental apparatus suggests that their diet was principally vegetarian: the shape of the teeth and the scratches etched in their surfaces when food was masticated matches those of modern-day plant eaters, not habitual carnivores.

We can say, therefore, that Darwin appears to have been correct when he suggested that Africa is the birthplace of humankind, and that one of the principal initial adaptations was the adoption of bipedal, as opposed to quadrupedal, locomotion. But he was incorrect when he speculated that the evolution of bipedal locomotion was linked to increased brain size and tool manufacture. There was no inextricable evolutionary package that propelled us along the road toward humanity. We became identifiably human in humanlike ways only relatively late in our evolutionary history. Until about 2.5 million years ago, we were essentially apes on two legs.

What Caused Our Ancestors to Stand Upright?

Anthropologists have never been reticent about offering explanations for why, some 6 million years ago, an African ape adopted a distinctly unusual mode of getting about. For instance, it has been suggested that bipedalism helped our ancestors avoid predators: by being able to stand up and see farther across the savannah than quadrupeds are able to, our bipedal ancestors could detect approaching danger sooner. Other scholars have pointed out that modern apes stand upright occasionally as a social display, usually as a warning to rivals. Could this have been the *origin* of habitual bipedal posture, as opposed, say, to an ability *conferred* by upright walking? Another popular explanation has been that the evolutionary pressure was associated with a shift in diet, with our apelike ancestor eating foods that apes traditionally did not, and that this required frequent upright walking and posture. Still another explanation has been the obvious one: that it allowed our ancestors to carry things and manipulate objects, in the now freed hands.

This last hypothesis was Darwin's preferred cause, as we saw earlier: specifically, that freed hands allowed hands to be busy in new ways, in making tools and weapons, for instance, as part of a newly developing mode of subsistence, that of hunting and gathering. As we said earlier, we now know that this is not correct. Freed hands undoubtedly allowed for hands that one day could fashion and use stone tools, but that "one day" was a very long way in the future: almost 4 million years, to be precise. We must point out, however, that we cannot rule out the use and role of tools made from more perishable materials such as wood, which are not preserved in the fossil record.

It is true, of course, that freed hands for carrying things—not manipulating things such as tools—might have been natural selection's edge in promoting bipedalism. But the freed-hands hypothesis exemplifies the danger of explaining the origin of a current structure in terms of its modern function. The enlarged human brain, after all, is unlikely to have evolved so that people might write symphonies or computer simulations. Something other than these activities spurred unusual brain-size evolution in our prehistory. Once so endowed, *Homo sapiens* could do things with this large cognitive capacity, too, other than the activity for which it evolved. The same may hold for freed hands and the corollary ability to carry things. It is possible that upright walking evolved because of the advantages of carrying or making things with emancipated hands. But it is equally possible that we can carry and make things so effectively because our hands became freed from a locomotor function for some entirely different reason or cause.

Although *Homo sapiens* is not the only primate to walk on two feet—for instance, chimpanzees and gibbons often use this form of posture in certain circumstances—no other primate does so habitually or with a free, striding gait. The rarity of our form of locomotion among primates—and among mammals as a whole—has encouraged the assumption that bipedal walking and running is energy inefficient and therefore unlikely, or at least difficult, to evolve. As a result, anthropologists have often sought "special"—that is, essentially human—explanations for the origin of bipedalism, such as the hypothesis about carrying things. As paleoanthropology has absorbed more mainstream biological thinking during relatively recent times, researchers have looked more and more toward ecological and environmental causes of human evolution, just as biologists would for any other animal, rather than to the special pleading of the past. This is a welcome development in the science, and it is an important statement about how we see ourselves as a species in the rest of the world of nature.

The nature of the evolutionary transformation that took place in the first member of the human family must, of course, have depended upon the mode of locomotion and posture of our common ancestor with the apes. The ancestor might have been a knuckle-walker, like the chimpanzee, a creature that for the most part walked quadrupedally while it was on terra firma, although it was capable of standing and walking upright for short distances and was extremely agile while aloft in the trees. When we compare the anatomy of a chimpanzee with our own anatomy, we can see that big shifts have occurred that are associated with the way we carry ourselves. For instance, our lower spine is more curved; we have a shorter, broader pelvis; our legs are relatively longer, our arms relatively shorter; we have a foot that acts as a flat platform, with the enlarged big toe brought in line with the other toes, not splayed outward; and the hole in the bottom of our skull through which the spinal cord passes—the foramen magnum—is positioned more toward the center, rather than toward the back.

These anatomical differences between us and our closest relative are quite substantial, profound in some cases, and this, too, has encouraged belief that the agent of natural selection that propelled the change must have been not only unusual in its power but also unusual in its nature. However, in recent years two lines of evidence have eroded this argument. First, it has become clear that until the advent of *Homo erectus*, almost 2 million years ago, our ancestors, though bipedal, sported bodies that were apelike in many ways. The shape of the trunk was more like an ape's, that is, somewhat conical rather than barrel-shaped; the neck was short and the waist virtually absent; the legs were relatively short and the arms relatively long. In other words, the evolutionary transformation from progenitor to the first member of the human family was not as dramatic, and therefore difficult, as was assumed.

The second line of evidence is ecological. Darwin was correct when he supposed that the human family arose in Africa. And he was probably correct to think that the initial push toward bipedalism came from a shrinking of forest cover as the climate cooled. But the widely held belief that our founding ancestor evolved as a novitiate out on the open savannah is almost certainly incorrect. We can say this because of new, hard evidence, a relative luxury in this line of inquiry about our origins. Two of the most recently discovered—and oldest—fossil species died, and therefore probably lived, in heavily wooded habitats. These species are *Ardipithecus ramidus,* which Tim White and his colleagues found in Ethiopia in 1993, which lived almost 4.5 million years ago; and *Australopithecus anamensis,* unearthed by Meave Leakey and her coworkers in northern Kenya in 1995, dated at a little more than 4 million years old. In both cases, the animal fossils found alongside the ancient human relics were of woodland and forest creatures, not savannah inhabitants. The beguiling notion of our intrepid forebears bravely striding out of dense forest and onto open grassland savannah, there to begin the evolutionary journey toward us, is likely to be more fiction than fact.

During the 1960s there was great emphasis on a "Man the Hunter" scenario for human origins, for several reasons. One had to do with the sociology of science, in that anthropologists at the time were learning volumes about the lifeways of modern hunter-gatherers. The technologically simple life of these people—specifically, the San people of the Kalahari Desert—seemed a good model for our ancestors' way of life. Another reason was purely sociological, in that it was a macho image, and many male anthropologists liked that. And, after all, hadn't Darwin said this was how we began? There was some science, too, of course. Human bipedalism was shown to be less efficient in terms of energy use than quadrupedalism in other mammals when running, and we are a good deal slower, too. (At a walking pace, however, humans are energy efficient.) Although this would be a disadvantage for a predator that kills prey in a foot race, it is an advantage to one that tracks prey over long distances and periods of time, striking perhaps through stealth rather than speed, because bipedal walking confers great long-term stamina. Even when the "Man the Hunter" image was modified into "Man the Scavenger"—less glamorous, but more biologically realistic—the endurance locomotion provided by bipedalism was still an advantage. It enabled the earliest humans to follow in the wake of migrating herds, for instance, opportunistically scavenging the carcasses of the unfortunate young and the infirm old—or so it was argued.

One little problem proved fatal to the scenario in either of its variations, however: not only do stone tools that are required for butchering carcasses not appear in the archeological record for as much as 3.5 million years *after* the first human species stood on its hind legs, but there is also no indication of regular meat eating by our earliest ancestors. Their teeth are the teeth of creatures that subsisted principally on plant foods, particularly *tough* foods, like unripe fruits and tubers.

Before the "Man the Hunter"/"Man the Scavenger" notion was felled by inconvenient facts from the archeological record, however, it came under attack from a different direction: women. It is perhaps not surprising that in the feminist decade of the 1970s the macho overtone of the hypothesis was challenged by women anthropologists. It wasn't male behavior that drove the evolutionary momentum toward novelty, they said. It was female behavior. According to the "Woman the Gatherer" hypothesis, plant foods, not meat, were major food items in our early ancestors' diet. (The archeological record was to prove them right on this point.) And it was plants, not meat, that were the focus for technological innovation and new social behaviors, claimed proponents of the hypothesis.

According to Adrienne Zihlman, of the University of California, Santa Cruz, and the principal champion of "Woman the Gatherer," in the social life of our earliest ancestors the focus was females and their offspring, who shared food gathered by the adults. Males were rather peripheral. This is very much what life is like for modern large primates, such as chimpanzees. The evolutionary novelty, argued Zihlman, was that because our ancestors lived in more open country than is typical for chimpanzees, the females had to travel greater distances during foraging. The envisaged technological innovation in this novel lifeway was the making and using of wooden tools to unearth underground tubers and the construction of rude containers for carrying food and infants. Hence, bipedalism would have provided a selective advantage.

The "Woman the Gatherer" hypothesis was more conservative than the "Man the Hunter" model, in that the first humans were viewed as being basically apelike rather than already essentially human. However, it assumed that the ability to carry and make things was the cutting edge of natural selection, not a benefit of freed hands. And, of course, it adduced open country living in the earliest humans, which fossil evidence later showed not to have been the case.

In 1981, Owen Lovejoy, an anatomist at Kent State University, made many headlines with his "Man the Provisioner" model, in which males gathered food and returned it to some kind of home base; there, the males shared the fruits of their efforts with females and offspring, specifically *his* female and *his* offspring. The model suggested that there was pair bonding and sexual fidelity with male–female couples. With their dietary requirements met by their males, the females would be free to reproduce with greater efficiency, that is, by having babies more frequently than they would if they had to spend time foraging. This, it was said, gave our early ancestors an evolutionary edge over other apes of the age. The system would work only if a male could be reasonably certain that the infants he was helping to raise were his—hence the need for pair bonding and sexual fidelity.

In other words, Lovejoy's model was the idealized Western nuclear family—present in our history, right from the beginning. Quite apart from the obvious sociological grounds for objection, Lovejoy's hypothesis was justifiably challenged on other fronts as well. The most pertinent was that if indeed the first humans were monogamous, as the model demands, then much of what biologists know about the life of primates had to be suspended. In monogamous large primates, such as gibbons, males and females are about the same body size. In other social systems—where males compete for the attention of females, for instance—males are always bigger than females, sometimes substantially so. How was it, Lovejoy's critics wanted to know, that in the earliest human species, males were much bigger than females—approaching the difference in size between males and females in modern gorillas, for instance—if they were lifelong partners?

Not all hypotheses about the origin of our bipedalism have dripped with sociological overtones. For instance, Peter Wheeler, of Liverpool John Moores University, England, saw bipedalism as a possible adaptation to heat stress. A bipedal posture reduced the area of body surface exposed to the sun while the individual was foraging, particularly during midday, he said. In this way, through physiological cooling, an upright posture minimized the heat stress the animal must cope with. The loss of body hair was, suggested Wheeler, a further way of increasing cooling efficiency, through the enhanced ability for copious sweating. Wheeler made a tight argument in terms of human physiology, but, again, the scenario placed our earliest ancestors out on the savannah, where coping with a blazing tropical sun would indeed have been an issue. But for creatures living in woodland, as the first human species apparently did, the issue is much less relevant.

One scientifically attractive explanation of the origin of bipedalism, proposed by Peter Rodman and Henry McHenry of the University of California at Davis in a 1980 publication, is based on the demands placed on an ape that found itself needing to survive in a novel habitat. It wasn't the savannah, as many hypotheses have assumed. Rodman and McHenry proposed a woodland habitat—as Darwin suggested—but one that was changing. According to the hypothesis, our earliest ancestral species evolved upright locomotion not as part of a change in the *nature* of the diet or social structure, but instead as a result of a change in the *distribution* of food. As global temperatures cooled—as they did around 6 million years ago—tropical forests in parts of Africa, particularly in East Africa for local geological reasons, became less a carpet cover and more a fragmented patchwork of trees. Our protohuman ape therefore found itself having to wander from patch to patch in order to forage for food, rather than move through an effectively continuous larder. Anything that increased the energy efficiency of that between-patch foraging would be evolutionarily favored. That "anything," suggested Rodman and McHenry, was getting up on the hind limbs and walking bipedally as a matter of continuous habit, not just occasional whimsy.

If, as reputable biologists have demonstrated, human bipedalism is less energy-efficient than the quadrupedalism of, say, a dog, how can an argument based on increased efficiency be credible? Simply because of the blindingly obvious fact that humans are not dogs. Our mode of locomotion may be less energy-efficient than our pet pooch's at running speed, but at walking speed we humans match our canine companion's energy efficiency quite well. More to the point, chimpanzees, with their shambling mode of knuckle walking, are not perfect quadrupeds. They are some 50 percent less energy-efficient than conventional quadrupeds when walking on the ground, whether they employ knuckle walking or move bipedally. If the common ancestor between humans and apes was anything like chimpanzees in terms of locomotion, then walking on two legs, not four, would have conferred an energy advantage on a descendent species that needed to cover large distances while foraging, as long as it could be done at a leisurely pace.

In other words, bipedalism was "an ape's way of living where an ape could not live," noted Rodman and McHenry. "It is not necessary to posit special reasons such as tools or carrying to explain the emergence of human bipedalism, although forelimbs free from locomotor function surely bestowed additional advantages to human walking."[12]

Already two decades since it was first proposed, Rodman and McHenry's energy-efficiency hypothesis is still the best explanatory bet on the bipedalism block. Not only is it plausible in itself—that is, in its mathematics—but it also has the merit of being distinctly biological in tone, as opposed to seeking recourse to special pleading involving some humanlike behavior. This being the case, it is appropriate to ask, What about the other apes? Why did the ancestors of chimpanzees and gorillas not also evolve a bipedal mode of locomotion and posture, if this was such a good way of coping with dispersed food sources?

According to Lynne Isbell, of Rutgers University, and Truman Young, of Fordham University, becoming bipedal is just one strategy for adapting to such ecological change. A second strategy is to reduce the distance that a group has to travel each day, which is achieved by having smaller social groups. (A large group requires more total food resources each day than a small group, and therefore must travel farther to harvest it.) It is a very flexible strategy, because when food is plentiful, social groups can be large, but when it is scarce, large groups can split up into smaller groups. This, say Isbell and Young, is precisely what you see in chimpanzees in Gabon. Gorillas, on the other hand, follow yet another strategy. When times are good, they feast on fruits. But when fruits are few, instead of splitting up into smaller groups, gorillas switch their diet to leaves.[13]

By placing our ancestor's evolution of bipedalism in the context of strategies by various ape species for adapting to environmental change, some 6 million years ago, we further demystify that evolutionary event. It wasn't the first step on an evolutionary journey, with *Homo sapiens* as its ultimate goal (blessed with freed hands, high intelligence, and technology). It just happened to turn out that way.

9

A Change of Body

THE biggest headline-grabbing discoveries in the realm of human origins research are the partial skeletons. There's Don Johanson's 3-million-year-old Lucy from Ethiopia, which he found in 1974, and the 1.6-million-year-old partial skeleton of a *Homo erectus* boy, unearthed by Richard Leakey's team on the west side of Lake Turkana, Kenya, a decade later. These are remarkable specimens, not just because their bones provide so many more scientific clues than are present in the more usual fragmentary discoveries, but also because they are impressive to look at. They are bold visual statements about our past.

Occasionally, however, a fragmentary, visually unimpressive fossil specimen can make bold statements about our past, too. Such is the case with the remains of an individual known to science simply as KNM-ER 1808, or, only slightly less prosaically, as "eighteen-oh-eight," a female *Homo erectus* who died about 1.7 million years ago. Kamoya Kimeu, Leakey's chief fossil finder, spotted her badly broken-up bones scattered over a very wide area on the east side of Lake Turkana in 1973, and organized their collection and shipment to the museum in Nairobi. There, they lay in boxes, little studied, and one of the least celebrated of human fossils to come from the Lake Turkana treasure trove of prehistory, but also one of the most evocative relics of the new kind of life that *Homo erectus* had introduced into the human story.

Not only were eighteen-oh-eight's bones broken into hundreds of pieces, but most of them were distorted by a strangely rough surface that made precise recognition even more difficult. That distortion, which effectively confined eighteen-oh-eight to relative anthropological obscurity, also, it turns out, speaks volumes about how she lived and died. With a little help from medical friends, Alan Walker, a longtime colleague of Leakey's, discovered that eighteen-oh-eight's bone pathology is precisely what happens when someone eats huge quantities of vitamin A.[1] Although many people are unaware of the fact, eating large quantities of this vitamin can be harmful—even, in the extreme, lethal. With vitamin A poisoning—or hypervitaminosis A, as the condition is called—a person experiences dizziness and nausea, the skin begins to break down, internal bleeding occurs, and the bones are affected. The tough, fibrous coating of the bones gets pulled off the bone, blood vessels get ripped, and blood pours into the newly created spaces and forms clots. If the individual lives long enough—a matter of only weeks—then the distorted, fibrous mass around the bones becomes ossified, forming matted bone. It is a painful death, and the person is extremely ill, quite incapable of taking care of herself or himself for very long.

What do eighteen-oh-eight's bones tell us about her and about *Homo erectus*? First, they say that *Homo erectus* was a meat eater, and that this particular individual had consumed the liver of a large carnivore; that is the most concentrated source of vitamin A on the African savannah, certainly sufficient to induce a fatal case of hypervitaminosis A. Second, we can tell that eighteen-oh-eight had extensive help from others; if they had not supplied her with food and water and kept predators at bay, she would have succumbed to death long before the disease had progressed as far as it so obviously did, leaving its signature in the surface of the bone. In short, the first bold statement eighteen-oh-eight makes about our past is that, with the arrival of *Homo erectus,* meat eating had become an important part of our ancestors' day-to-day subsistence, for the first time in human prehistory. And second, a degree of sociality—collective care and support—had developed that is absent in all large primates today, and almost certainly in all human species prior to *Homo erectus.*

A dramatic change had taken place in the human lineage, one that was expressed in many more ways than meat eating and sociality. It was reflected in the size and shape of the body of *Homo erectus* and in a slew of associated behaviors that were different from those of its forebears. It could also be seen in an increase in brain power and the innovative behaviors this allowed; and almost certainly it was manifested in the ability—albeit probably limited—to produce spoken language (the next chapter).

As we said, there was a new kind of animal under the sun, when the Mojokerto child's species, *Homo erectus,* came on the scene—not yet fully human, but no longer essentially ape.

Bodies of Evidence

Because skeletons—or even partial skeletons—are extremely rare in the human fossil record prior to Neanderthal times (beginning about 150,000 years ago), anthropologists must play the role of detectives when they ask, What did the earliest members of the human family look like, specifically in their overall stature and body proportions? All that anthropologists have to work with is a motley collection of bones, mostly broken and incomplete petrified bones. But by measuring the length and thickness of certain bones, particularly the leg bones, and by determining the size of certain joints, such as the knee and hip, we can "reconstruct" the bodies of long-extinct species. And from these reconstructed bodies it is then possible to reconstruct past lives (in outline, at any rate).

The key evolutionary innovation that defines the human family, and which was the basis of nature's experiment through adaptive radiation, was habitual bipedal posture and locomotion. From what can be said of all the species prior to *Homo erectus* times, it looks very strongly as if these creatures, while being well adapted to walking upright while on terra firma, were also adept tree climbers. The hand bones, which were curved in a way that is reminiscent of those of chimpanzees, seem to speak of arboreality as a still important part of the locomotor repertoire of these creatures.

If you were able to stare these individuals (the australopithecines) in the face, you would very quickly come to the descriptive phrase "bipedal apes." Although they lacked the large canine teeth so characteristic of apes, their small crania housed brains little bigger than those of apes, and their faces protruded forward, like a chimpanzee's face. There were variations on these themes, of course, principally around the teeth and jaws: while the cheek teeth in all these early human species were flatter and bigger than those in apes, and the front teeth smaller, in some of them these features were greatly exaggerated. Those species with super-big cheek teeth and jaws, which look as if they were adaptations to grinding tough foods, earned them the description "robust australopithecines"; the ones with relatively smaller cheek teeth and jaws are known as "gracile australopithecines."

Although the robust and gracile species would have appeared quite distinct in their faces and heads, their bodies were quite similar. As was noted in the previous chapter, the shape of the trunk was distinctly apelike, that is, the trunk was somewhat conical rather than barrel-shaped; the neck was short and the waist virtually absent; the legs were relatively short and the arms relatively long. While recognizing that we are lumping different species togther—which is not too unreasonable, because they are all rather similar—we can get an average weight for the pre-*erectus* australopithecine species of about 84 pounds and a stature of 4.2 feet. Although the average hides a small amount of variation *among* different species, much more important is the variation *within* species—that is, the difference between males and females. The average male australopithecine weighed in at around 94 pounds and stood 4.6 feet tall; females were a diminutive 70 pounds and 3.75 feet.

This degree of so-called sexual dimorphism in body size is very characteristic in primate species with a certain kind of social structure: namely, where there is keen competition among the males for sexual access to the females. It's a blunt truth of the primate social world that the big guys usually win, so that some of the males will have the opportunity to produce many offspring, while others will be much less fortunate. This degree of sexual dimorphism also implies that females, with their young, range over smaller territories while foraging than do their males.[2] In other words, the earliest members of the human family were apelike not only in much of their anatomy but also in their social behavior.

With the advent of *Homo erectus,* the story changes—dramatically. We are going to do some averaging again, for simplicity, but we should especially note that although some anthropologists view *Homo erectus* as appearing close to 2 million years ago and disappearing soon after half a million years ago, others believe that there were two very similar species, *Homo erectus* and *Homo ergaster,* at this time. To keep it simple here, we will speak only of *Homo erectus.* The first dramatic shift to be noted is in average size (that is, the average male and female), 128 pounds and 5.7 feet; that's an increase of more than 50 percent over the average for australopithecine species. There is a thriving industry among biologists in the study of what are called life history factors, such as longevity, age at maturity, age at weaning, and so on. Body size has a strong influence on these factors. For instance, the average *Homo erectus* probably lived six years longer than the average australopithecine—that is, 50 years as against 44. *Homo erectus* individuals matured later than australopithecines, and started breeding later, too; their gestation was longer, and they produced bigger babies. Being bigger made *Homo erectus* less vulnerable to predators.

The increase in average body size was therefore a major landmark in human prehistory, and it brought many behavioral shifts in life history factors. Even more dramatic, however, is what occurred with sexual dimorphism in body size. Although the average *Homo erectus* male was close to 50 percent bigger than the average australopithecine male, the difference in females was much more marked: *Homo erectus* females weighed almost 70 percent more than australopithecine females, 117 pounds as against 70 pounds. The difference in body size between male and female *Homo erectus*—138 pounds (and 6 feet tall) compared with 117 pounds (5.3 feet tall)—is therefore much smaller than in earlier species. (There had been some reduction in body size dimorphism from the earliest to the latest pre-*erectus* species, but nothing like the quantum shift that happened with *erectus* itself.) What drove the increase in the body size of *Homo erectus*? And why did sexual dimorphism decrease so substantially?

There are several possible answers to the first question, and they need not be mutually exclusive.[3] Diet and a new mode of subsistence almost certainly played a part, probably linked to an increase in brain size. There is, as we'll see, plentiful evidence that *Homo erectus* individuals ate far more meat on a regular basis than earlier human species did, and perhaps had developed a primitive hunting and gathering way of life. Such a diet is richer and packs more energy than a more completely herbivorous regimen; with more energy, building bigger bodies is possible, as is building bigger brains. Being bigger is an advantage in a species that eats a significant amount of meat, because it allows for a more extensive home range—a must for a carnivore. More on this later. If there was increased aggression between neighboring *Homo erectus* groups, populations with bigger bodies would fare better in the long run. There is in fact no evidence for boosted aggression, but this does not mean it was necessarily absent.

Why the reduced sexual dimorphism? This was probably a complex—rather than a single-issue—change, too. The simplest answer is that competition among males was greatly reduced, perhaps because the males in *Homo erectus* social groups were related to one another, as is the case in chimpanzee groups and in technologically primitive human social systems. Reduced rivalry between males leads to less evolutionary pressure for males to be big. But the fact that the dimorphism got smaller because females increased in size markedly more than did the males was probably a signal of other factors, too, such as the need to nurture larger brains in babies, *in utero* and out.

Whatever the cause(s) of the change in overall body size and in sexual dimorphism, there can be no doubt that the daily social life of *Homo erectus* was very different from anything that had gone before, and was much more like what came afterward.

When Herbivore Turns Carnivore

Modern large primates—excluding humans—are essentially plant eaters, subsisting mainly on leaves or fruit or both. But chimps and gorillas, for instance, also eat insects, grubs, and small animals, such as lizards or mice, when they come across them; and chimps actively go in search of ants and termites when they are "in season," and often use tools to harvest them. Chimps also occasionally engage in deliberate, collective hunting of monkeys, which they apparently prize greatly but digest very poorly. Baboons, too, from time to time go into hunting mode, and chase and capture small antelope. Again, the meat is obviously prized in the social group, with individuals variously exerting muscles and wiles to get access to it. But like chimps, baboons gain few nutrients from meat, because there is not much of it, and it is very poorly digested. This kind of mixed—but heavily vegetarian—diet very probably is how the early, pre-*erectus* human species subsisted. This notion is backed up by the clues to the australopithecine diet that Alan Walker has found when looking at the surface of fossil teeth, using scanning electron microscopy. The microwear pattern on the teeth of robust australopithecines, for instance, is very similar to the pattern on the teeth of modern species that eat a lot of tough fruits, and perhaps tubers; similarly, the gracile species appear to have subsisted on softer plant foods.

The teeth of *Homo erectus* tell an entirely different story. Rather than being relatively smooth, as in the australopithecine teeth, the surfaces of *Homo erectus* teeth are pitted and gouged, a pattern very reminiscent of what you see in the teeth of meat- and bone-eating carnivores, such as hyenas, says Walker. "All this information indicated that [early human species] had made an important dietary transition from a more plant-based to a more animal-based diet," he has concluded.[4] Such a shift is a relatively rare evolutionary event, perhaps because of the demands it places on many aspects of an animal's life.[5] For a meat eater, obtaining food (that is, catching prey) and processing it (that is, preparing it for ingestion and then digesting it) are very different challenges from those faced by plant eaters. And being on top of the energy pyramid—with plants capturing the sun's energy at the bottom, herbivores coming next, carnivores last—has consequences, too: a given area of land can sustain a hundred times as many herbivores as carnivores. If, in an evolutionary context, you change from being an herbivore to being a carnivore, you have to find a way of reducing your population; otherwise the evolutionary venture will fail.

Homo erectus shows every sign of having made that transition. In the challenge of obtaining prey, for example, the species shows two evolutionary novelties that are consistent with a more carnivorous way of life. One is anatomical, the other behavioral.

Much of what anthropologists know about the detailed anatomy of *Homo erectus* comes from the remarkable partial skeleton that Alan Walker helped unearth, on the western shore of Lake Turkana, in 1984. The skeleton—that of a boy who died at the age of about 12 years, some 1.6 million years ago—shows not only that *Homo erectus* was bigger than previous human species, but also that its body proportions had changed. The trunk was human-shaped, not the pot-belly form of apes; the legs were relatively long compared with the arms; and the lower legs and lower arms were longer than the upper legs and upper arms. All of these represent adaptations for running, argues Walker. Not that *Homo erectus* should be thought of as a cheetah of the human family; humans are relatively slow runners in the world of animals. But there is no doubt that *Homo erectus* was a more capable runner than any previous human species, an ability that would be important in a would-be meat eater.

Not all hunters of prey rely on speed, of course. Stealth and cooperative hunting, for example, are equally potent weapons. The degree of sociability of *Homo erectus*—as revealed by the nurturing that the unfortunate eighteen-oh-eight received during her dying weeks—was surely adequate for a species that obtained prey through cooperative hunting. It is also worth pointing out that most carnivores hunt when they have to and scavenge when they can. That is true of modern hunter-gatherers and was probably the case for *Homo erectus*. One of the more florid archeological debates of recent times centered on this very question: When, and to what degree, did our ancestors become hunters?

As we mentioned earlier, there emerged very strongly in the 1960s the notion that hunting had played a key part in human evolution. For instance, two prominent scholars proclaimed the following at a landmark conference in 1966, titled "Man the Hunter": "To assert the biological unity of mankind is to affirm the importance of the hunting way of life." Human hunting is special, they suggested, because it is "based on division of labor and is a social and technical adaptation quite different from that of other animals."[6] In more colorful terms, Robert Ardrey wrote that "man is man, and not a chimpanzee, because for millions upon millions of years we killed for a living."[7] The cogency of the hunting hypothesis is easy to understand. It offered a plausible explanation of the key behavioral and anatomical differences between humans and apes. And anthropologists had living analogues of this way of life, among technologically primitive hunter-gatherers of the modern world. No one was suggesting that the !Kung San of the Kalahari, for instance, were Stone Age people in anything but their mode of subsistence. But their way of life did seem ageless, as if it stretched back to a time when our ancestors differentiated themselves from the world of apes.

With this mind-set, archeologists fell into the easy habit of viewing collections of bones and stones they found in the prehistoric record as timeless signatures of a hunting and gathering habit. "It seemed a very attractive interpretation," says Rick Potts, an archeologist at the Smithsonian Institution in Washington. "The home-base, food-sharing hypothesis integrates so many aspects of human behavior and social life that are important to anthropologists—reciprocity systems, exchange, kinship, subsistence, division of labor, and language. Seeing what appeared to be elements of the hunting-and-gathering way of life in the record, in the bones and stones, archeologists inferred that the rest followed. It was a very complete picture."[8]

Lewis Binford, the brilliant and combative archeologist at Southern Methodist University, Dallas, challenged this view with a series of papers and books during the 1970s and 1980s. Something of a pitched battle ensued between Binford on one hand and Glynn Isaac of Harvard University on the other. Perhaps in an overreaction to the uncritical thinking of earlier anthropologists, Binford was soon arguing that there is no evidence for any form of hunting—only for marginal scavenging, as he called it—in the human record until very recent times.

As mainstream archeologists view it these days, the truth almost certainly lies somewhere between the two extremes. Although the occurrence of animal bones and stone tools at a place on an ancient landscape probably does not indicate the complexity of a hunting and gathering social life that the image so readily conjures up, it is a reasonable inference that they represent more than casual debris of marginal scavengers who simply scraped remnants of meat from bones long discarded by more competent carnivores. The presence of grooves on the animal bones—the result of a sharp stone having been used to remove meat—at ancient archeological sites supports the conclusion that from *Homo erectus* times onward, our ancestors were accomplished meat procurers and eaters. And in their social life and economic life they probably had more of a complementary arrangement of plant gatherers and meat procurers than was implied by Binford's "marginal scavenger" characterization.

Signatures of use on the edges of stone tools from a 1.5-million-year-old archeological site in northern Kenya provide intriguing clues. The tools, which were analyzed under the microscope by Lawrence Keeley, of the University of Illinois, and Nicholas Toth, of Indiana University, had been used, variously, for cutting meat, whittling wood, and cutting some kind of soft plant tissue. The site has been interpreted as a temporary living space for a small *Homo erectus* group, a soft, sandy bank on the bend of a small stream, with remnants of animal bones—including one that had apparently been smashed, which would have given access to nutritious marrow—scattered around. These fragments of evidence are strongly indicative of a social and economic complexity that goes far beyond what is seen in modern large primates, even if it does not match that of modern hunter-gatherers.

The use of sharp stone tools as a means of cutting through tough hide was, of course, a necessary innovation for a primate that wants to eat meat but lacks sharp canines to do the job of getting access to it. The earliest known such implements date back to about 2.5 million years ago, from sites in Ethiopia and Kenya, and are very simple, not much more than the product of knocking two pebbles together. More complex tools appeared during *Homo erectus* times, about which we will say more later. Here, the point is that, having dealt with the challenge of procuring meat, our ancestors faced a second challenge, processing the food, of which making and using tools was the first aspect.

A second aspect was digesting the food. Plant eaters have very bulky intestines, principally because of the lengthy process of digesting the cellulose that is so abundant in plant tissue. This is why apes are pot-bellied and why, presumably, pre-*erectus* humans were, too. If *Homo erectus* had indeed become a habitual meat eater, then its intestines would have changed, becoming less bulky overall but with more elaborate small intestines. It is hard to imagine how the change in the small intestine might show up in the prehistoric record, but the overall size of the gut does. The Turkana boy's skeleton reveals that this indeed was the case: his barrel-shaped thorax, well-defined waist, and narrow hips simply could not have accommodated the bulky intestine of a committed plant eater, but they were appropriate for the smaller gut of a meat eater.

The last signal of *Homo erectus*'s meat-eating habit relates to its position on top of the energy pyramid. There are always far fewer predators than there are prey, for obvious reasons; and the bigger the predator, the fewer of them any given area of land can support. This rule of ecology is captured in the title of a book by Paul Colinvaux, of Ohio State University: *Why Big, Fierce Animals Are Rare.*[9] *Homo erectus* may or may not have been fierce; we have no way of finding out. But it should have been rare. More accurately, its population density must have been lower than that of its ancestral species, a presumed plant eater. An alternative evolutionary strategy for the herbivore-to-carnivore transition in this context would have been to reduce body size: The ecological equation has to do with how much biomass an area can sustain, not just with the number of individuals. As we know, *Homo erectus* increased its body size, not the reverse, so the biomass problem would have been even more pressing. Being bigger may have been a sensible evolutionary strategy for a would-be hunter that is not particularly well equipped to confront other carnivores, equipped as they are with sharp teeth and claws. In any case, having increased its body size, *Homo erectus* had no choice but to reduce its population density, which could be achieved by expanding the home range.

Anthropologists have known for a long time that *Homo erectus* did expand its home range: the fossils and stone tools that have been collected in Europe and Asia for a century clearly attest to that. But until we produced our dates for the Javan fossils, and before the discovery in 1991 of a *Homo erectus* jaw in Diminisi in the former Soviet Georgia, which may be as old as 1.7 million years, and the discovery of two crania from the same site and of the same age in the summer of 1999, anthropologists believed that *Homo erectus* had expanded its range beyond Africa almost a million years *after* it first evolved—that is, not until about a million years ago. Why *Homo erectus* played such a waiting game was a major puzzle. Perhaps the biological imperative to thin out its population was feeble, because it was an indifferent carnivore? Perhaps, as some people suggested, it needed a technological assist? The ancient dates for the Mojokerto child and the Sangiran cranium, and for the Diminisi fossils, clearly show that no technological assist was needed, and that, driven by a newly evolved predatory habit, the biological imperative to reduce population density by expanding the home range was loud and clear, right from the beginning.

10

A Change of Mind

ANTHROPOLOGISTS have long been obsessed with the two-plus pounds of gray matter we sport in our heads. Although bipedalism is the defining characteristic of the human family, the large brain of *Homo sapiens*—three times the size of the brain of an ape with an equivalent body size—has come to define what we, as a species, have become. Our intelligence, creativity, reflective consciousness, ability to use language—these collectively represent what we mean by being human, and collectively separate us from the rest of the world of nature, at least in degree if not always in kind. Counterintuitively, exactly what selective pressure propelled such a dramatic increase in brain size, and more particularly in brain power, is not easily identified. The "obvious" answers—such as "for increased technological skill" or "for enhanced communication"—may have contributed to our ballooning brain through evolutionary time, but they fail as complete explanations. By looking at brain evolution in *Homo erectus* and at the behaviors that accompanied it, we can reach a sense of what might have been driving this important facet of our prehistory.

Darwin, remember, believed that brain expansion began at the outset of our evolution, part of an evolutionary package that linked bipedalism, technology, and behavior. Until relatively recently, this was the accepted view. But as the fossil and archeological records improved—including the finding of a profoundly evocative trail of footprints at Laetoli, Tanzania, made by our ancestors some 3.75 million years ago—our understanding changed; we now know that Darwin had the wrong idea of how our history unfolded. The founding ancestor of the human family first stood upright some 5 million years ago. The first record of stone tools followed 2.5 million years later: they date to 2.5 million years ago. This coincides with the earliest known fossil remnants of our own genus, *Homo.* But because the fossil record is so sparse for our lineage at this time, it isn't possible to say whether or not the brain had already begun to expand 2.5 million years ago. The first evidence that allows this question to be asked—from crania close to 2 million years old—does show a substantial brain size increase over that in australopithecines: a 50 percent increase, to some 660 cubic centimeters. It is a reasonable guess that when anthropologists find a 2.5-million-year-old *Homo* cranium that is complete enough to allow brain volume to be measured, it will show that it once contained an enlarged brain. The guess is based on the intuition that first tool making and first brain expansion are linked, although not everyone would agree with this suggestion.

While the earliest history (that is, 2.5 to 2 million years ago) of the genus *Homo* still lies waiting to be discovered somewhere in Africa, the next known stage is fragmentary at best, and confusing. Prior to, and in some form ancestral to, *Homo erectus* there lived at least one, and possibly two, species of *Homo:* they have been named *Homo habilis* and *Homo rudolfensis.* Avoiding the debate over how many species of *Homo* lived at this time, what they were, and what their fate was, we will simply note that, although their brains had enlarged over that of australopithecines, their average body size, and the difference in size between males and females, were not much different. Although the earliest *Homo* was obviously better endowed cognitively than its ancestors, having the ability to fashion simple stone tools, there is no indication of the suite of anatomical and behavioral changes that speaks of the kind of major life change we speculate about in *Homo erectus.*

The life of *Homo erectus* went through a portentous transition not just because it had a big brain on its shoulders, but because its body also changed in size and shape, a fact whose significance has only recently been recognized. In fact, although the brain in early *Homo erectus* was almost 50 percent bigger than in *Homo habilis*—an average size of 950 cubic centimeters as compared with 660—the *relative* brain size remained much the same, because the body of *Homo erectus* was so much bigger, too. Some, perhaps all, of the extra brain machinery must have been required for the basic operation of a bigger body—just as elephants have bigger brains than we do but are not smarter. It is also possible, however, that increased size did confer some extra cognitive edge.

The bigger brain of *Homo erectus* had several consequences, some of a basic biological nature, others more in the realm of important cognitive skills, such as innovation and making sense of the world, perhaps through a nascent spoken language.

Big Brains and Childhood Time

Birth is a hazardous time for all mammals, both the process itself and the immediate aftermath, a time of extreme vulnerability for a newborn in a harsh world; predators, and sometimes an inclement climate, threaten to bring a young life to an abrupt end. For humans, birth is doubly dangerous. First, our enlarged brain makes birthing itself especially hazardous; the percentage of babies who become stuck or brain-damaged during delivery is far higher than in other species. Second, the amount of brain growth that is required after birth is far greater than in other species, so that human neonates are particularly helpless, and require extensive nurturing and protection. The flip side of that nurturing overload—extended childhood, as it is known—is that it is an opportunity for learning social and practical skills before the individual has to make his or her way in the world.

For all species, building brains is an extremely expensive metabolic proposition. Primate mothers pour an inordinate amount of energy into their growing fetus, a large slice of which goes toward brain growth. Once the infant is born, the rate of growth of its brain drops substantially: the mother simply doesn't have the metabolic resources to maintain such a frantic process, once the offspring is outside her body. In African apes, for instance, the neonate is born with a brain about half the adult size, which it attains within a year or so, but through a lowered rate of growth.

Humans, however, are different. The separate demands of a pelvis that is mechanically efficient for bipedal locomotion and at the same time can deliver a neonate with a large brain puts a limit on the size of brain that babies can be born with: about 385 cc. If human babies followed the ape growth pattern of doubling brain size after birth, we would finish up with brains of about 770 cc, which is less than 60 percent of the actual size of the adult human brain, some 1350 cc. In order to achieve the more than tripling in brain size that occurs in between birth and adulthood, evolution had to come up with a few tricks. One was to maintain the high fetal rate of brain growth for a year after birth: effectively, therefore, human gestation is 21 months, not the 9 that go on inside the womb. Another was the elaboration of social cooperation required to sustain an extended childhood.

The question here is this: Did *Homo erectus* follow the ape pattern of simply doubling its brain size after birth, thus making a nurturing extended childhood unnecessary, or was it already humanlike to some degree? With recourse to what can be learned from the pelvis of the Turkana boy, plus a little arithmetic, this question about an ancient human behavior can be answered. Let's assume, for the moment, that the pelvis of a female *Homo erectus* could accommodate a baby with the same head size as a modern woman's can. *Homo erectus* babies would therefore be born with brains of about 385 cc. If the ape pattern of brain growth followed—that is, doubling in size—then the maximum achievable adult brain size would be 770 cc. This is less than the average adult brain size found in early *Homo erectus* fossils, by some 90 cc. Which means that in the best of circumstances—that is, if early females had a modern-sized pelvic opening—the ape pattern of brain growth would fail to produce an adult brain of the required size. There must have been a shift toward the human pattern, even in early *Homo erectus.*

What if the pelvis of early *Homo erectus* females could not negotiate a human-sized neonate? Our calculation would be different. Although the Turkana boy is of course male, it is possible to calculate from his very well preserved pelvis what a female's birthing anatomy would look like, in terms of size. The answer, Alan Walker calculated, is that the limit on head size for early *Homo erectus* babies was about 275 cc. Only a human growth pattern—a tripling of neonatal brain size—would achieve the known adult size. A humanlike extended childhood can therefore be added to the list of behaviors that made *Homo erectus* a much more humanlike species than anything that had arisen earlier in the human family.

Brains and Technology

The notion that technology—that is, the need to use and make tools—was *an* engine of brain expansion, if not *the* engine, has a simple logic to it, and a long history. Again, Darwin had linked the two, even if he was wrong about the timing of this evolutionary tango. And in the 1940s and 1950s, the idea that not only did humans make tools but that, effectively, tools had made humans was a very popular theory, encapsulated in a 1949 book by the British anthropologist Kenneth Oakley, titled *Man the Tool Maker.* Indeed, until Jane Goodall's observations of chimpanzees and their exploits in making and using tools of various kinds, tool making and using was considered a defining characteristic of humanity.

The attractive simple logic of the technology/brain-growth link is that the two seem to have arrived simultaneously in human prehistory, beginning about 2.5 million years ago. As this is about half a million years prior to the evolution of *Homo erectus,* we must assume that its precursor species, *Homo habilis,* was also a toolmaker. (Although it is most parsimonious to argue that only members of the genus *Homo* were toolmakers, at least one anthropologist argues that some australopithecines may also have had the ability.)[1] Presumably, *Homo habilis* was also a meat eater, but probably not to the same extent or in the same social and economic context as *Homo erectus.*

Archeologists have built careers on classifying different types of stone tools found in the archeological record and trying to understand what they were used for. Here, we are interested in what the tools tell us about the mind of *Homo erectus.* The first, and most striking, conclusion about tool making in prehistory is that nothing changes very quickly. The first tools—pebble tools, as we said—consist of perhaps half a dozen types, such as scrapers, choppers, discoids, and the sharp flakes that are produced in the manufacture of these tools. Simple though they may be, these tools can be used for a great variety of tasks, in the right hands attached to the right ("right" in the sense of "correct") brain.

This so-called Oldowan tool technology is pretty much all there is to be found in the archeological record for about a million years. At that point, the range of tools being produced increases, and some of them are quite elaborate, such as the Acheulean hand axe mentioned in chapter 6. But, once established, the Acheulean technology changes little for a million years. The appearance at that point of much more extensive and more delicate tools coincides with the appearance of more advanced forms of *Homo,* the presumed maker of the new tools.

The archeologist Nick Toth specializes in making ancient stone tools as a path to understanding something about them, such as the necessary mode of manufacture, and gaining an insight into the skills needed to make them. Making an Acheulean hand axe, he says, requires much more mental planning, much more systematic manual implementation, than is involved in making an Oldowan chopper, for instance. In other words, makers of the Acheulean tools had mental models in their heads of what they intended to produce, and skilled hands to carry out the task. What of Oldowan tools? How much skill is required there? Were the earliest toolmakers doing something that was beyond the cognitive capability of apes? Or were they merely bipedal apes who were applying apelike cognitive skills to non-apelike activities?

A decade ago, Thomas Wynn, an archeologist, and William McGrew, a primatologist, threw down a challenge. In a paper they titled "An Ape's View of the Oldowan," they argued that tool making was a matter of spatial skills. Let's see, they wrote, what spatial skills chimpanzees have, compared with the skills required to manufacture Oldowan tools. In the context of stone tool making, they proposed that the skills consisted of landing a blow where it was intended, understanding that a target stone had different parts, and being able to deliver blows in the correct order. "All the spatial concepts for Oldowan tools can be found in the minds of apes [in other contexts]," they concluded. "Indeed, the spatial competence [we observed in chimpanzees] is probably true of all great apes and does not make Oldowan tool-makers unique."[2]

Toth decided to put this assertion to his own practical test, and joined with Sue Savage-Rumbaugh, of the Yerkes Language Research Center, in the venture. Savage-Rumbaugh had been working since 1980 with a pygmy chimpanzee named Kanzi, testing his ability to understand symbols. Toth thought Kanzi would be a good candidate to determine whether indeed the Oldowan technology is within an ape's cognitive grasp. Toth and colleagues of Savage-Rumbaugh's patiently showed Kanzi how they made sharp flakes by knocking one rock against another. Kanzi quickly developed the ability to discriminate between flakes that were sharp and ones that were not—he would use the sharp flake to cut a string and gain access to food in a box—but he never learned to make flakes efficiently, not of the kind that Toth routinely makes or that are found in the fossil record.

Kanzi did discover that he could make flakes—albeit randomly shaped—by throwing a rock against a hard surface, which demonstrated that he understood the goal of the operation, knew of the utility of the flakes once obtained, and could invent ways of getting what he wanted. But he was not what you would call a stone tool maker. Although they look primitive, the Oldowan tools are the product of a demanding process known as hard-hammer percussion. The process is as follows. First, the target rock, or core, must have an acute edge (one with an angle of less than ninety degrees). Second, the core must be struck with a sharp, glancing blow, hitting about half an inch from the edge. And third, the blow must be directed through an area of high mass, such as a ridge or a bulge. Do this skillfully and you can produce long, sharp flakes, and choppers and scrapers and discoids, all of which have the appearance of great simplicity.

As Toth points out, it is the process, not the product, that reveals the complexity of Oldowan tool making. The same can be said of Acheulean tool making. (Wynn, incidentally, does concede that there is something "very human" about Acheulean axes.) It has to be said, however, that until about 200,000 years ago stone tool assemblages have the appearance of being the product of relatively simple minds, particularly given the mind-numbing lack of innovation that stretched for hundreds of thousands of years.

What does all this tell us about the mind of *Homo erectus*? First, it suggests that *Homo erectus* was endowed with cognitive skills that exceeded those of modern apes, at least in the tool-making realm. But when we consider the cognitive skills of chimpanzees that are revealed under laboratory conditions—the ability to do complex, analytical problems, for instance—we know that the mind of *Homo erectus* must have been much more complex too. Just how complex is difficult to infer from looking at the tools produced. Superficially, at any rate, we can venture to say that even if the need for technology was part of what initiated brain expansion in our ancestors, it was probably not the only, or even major, factor sustaining it.

That said, we are still left with a puzzle. Oldowan technology first appeared some 2.5 million years ago, the presumed handiwork of *Homo habilis*. *Homo erectus* evolved soon after 2 million years ago, and it, too, appears to have manufactured Oldowan-like tools. Then, 1.5 million years ago, the Acheulean assemblage appears for the first time. It would be much neater, for writing the story of human prehistory, if the Acheulean assemblage and *Homo erectus* arrived on the scene simultaneously. Is the half-million-year gap that separates their appearance simply a matter of an incomplete—or misread—record, as it was with the dating of the Javan fossils? Or must we seek some as yet elusive explanation for why members of a species suddenly invented a new technology half a million years after the species' first appearance and then invented essentially nothing new in the technological realm for another million years?

Brain and Language

If our large brain defines our species for what it has become, language is surely its quintessence. If we are obsessed with how our brains have become so powerful, we are positively enraptured by what can be woven on the loom of language. Ours is a world of words, from the mundane orbit of practical affairs to the intellectual and spiritual sphere of abstraction, mythology, and religion. Language liberates us, and constrains us, by imbuing the myriad elements of life with meaning while also leaving some of the most important things in our lives—like love—beyond language, at least beyond the language of explanation. Language can stir our emotions—sadness, happiness, love, hatred—and it allows us to communicate the truth or to deceive with lies. Through language we can express individuality or demand collective loyalty. Quite simply, language is our medium, the bedrock of our humanity, so that a world without words is unimaginable to most of us.

Thomas Henry Huxley, Darwin's friend and champion, opined the following about his judgment of the importance of human language, in his 1863 book, *Man's Place in Nature:* "No one is more strongly convinced than I am of the vastness of the gulf between . . . man and the brutes . . . for he alone possesses the marvelous endowment of intelligible and rational speech [and] . . . stands raised upon it as on a mountain top, far above the level of his humble fellows, and transfigured from his grosser nature by reflecting, here and there, a ray from the infinite source of truth." Biologists and anthropologists today continue to be equally impressed with this precious mantle that so separates us from the rest of nature. And there is no topic in anthropology that provokes such a range of opinions as the evolutionary origin of language—when did it happen, and what honed it in our ancestors' minds? And there is no topic on which—appropriately enough—so much has been written, the mass spilling of words in earnest pursuit of why we use words at all.[3]

To Alfred Russel Wallace, our possession of spoken language was so far beyond the realm of what might ordinarily be conferred by natural selection (which he co-invented with Darwin) that he considered it the gift of the supernatural. In a more grounded way, David Premack, of the University of Pennsylvania, expresses a similar sentiment: "Human language is an embarrassment for evolutionary theory because it is vastly more powerful than one can account for in terms of selective fitness."[4] But Steven Pinker, a linguist at the Massachusetts Institute of Technology, is not so easily embarrassed. In his book *The Language Instinct,* he says that "People know how to talk in more or less the sense that spiders know how to spin webs."[5] In other words, humans obtained language the old-fashioned way: it evolved by natural selection. "It is fruitful to consider language as an evolutionary adaptation, like the eye," he says, "its major parts designed to carry out important functions."[6]

The question for anthropologists is, What are—or were—those functions? A related question is the following: Is the neurological machinery that elaborates spoken language in the brains of *Homo sapiens* built upon cognitive apparatus in the brains of our ancestors that was related in some distant way to language, such as communication or a facility for abstraction?

The intellectual territory into which we are wandering is huge, and beyond the scope of this book. Having simply recognized some of the major elements of the territory, we wish to concentrate on a question more directly related to the theme of this book. When the Mojokerto child died at the tender age of six, was it already developing a syntax and a vocabulary of a means of communication we would recognize—however distantly—as spoken language? Did *Homo erectus* possess the loom of language, however primitive, or was the species no more than a bipedal ape in its mode of communication?

Social Brains

Why did language evolve? From the point of view of natural selection, its usefulness for communication is the most obvious answer. Human spoken language is unprecedented among animals, both for information content and rate of transmission. In the modern world, where information is the most powerful currency of action and ideas, we constantly use language to communicate. And it is easy to imagine how very useful language would have been to hunter-gatherers for organization and planning. Seductive as it is, however, communication has become less favored as the agent of natural selection for language origins.

For instance, Harry Jerison, a neurologist at the University of California, Los Angeles, who has made a special study of brain evolution, has argued that "the role of language in communication first evolved as a side effect in the construction of reality." Different senses are important to different species in their mental construction of their realities. In amphibians, for instance, vision provides the principal element of that world; for reptiles, an acute sense of smell. For the earliest mammals, hearing was additionally important; and in primates, a mélange of sensory input creates a complete mental model of external reality. And, suggests Jerison, language represents a further tool that the human mind uses for creating the human world. "We can think of language as being merely an expression of another neural contribution to the construction of mental imagery," he says.

Through language, or, more precisely, through reflective thought and imagery, the human mind creates an internal model of the world that is uniquely capable of representing—and coping with—complex practical and social challenges. Inner thought, not outer communication, was the facility upon which natural selection worked, argues Jerison. Language was its medium—and, at the same time, it is an efficient tool for communication. In this context, it is impossible to separate the evolution of language from the evolution of introspective consciousness, a human facility that philosophers have wrestled with for millennia. Both language and consciousness, anthropologists are coming to believe, evolved in the increasingly intense nexus of our ancestors' social and economic lives. Individuals need to be able to understand, and predict, the behavior of others in their group, and the most efficient way to do that is to create a complex model of social interaction, partly by being intensely aware of one's own behaviors and motivations. Language and introspective consciousness can combine to do that. This has come to be known as the social brain hypothesis.[7]

Many Lines of Evidence, Different Conclusions

When did spoken language evolve in our ancestors? Was it, for instance, a slow, gradual buildup of language capability, beginning in primitive form with the origin of our genus, *Homo habilis,* and culminating in fully modern language, with *Homo sapiens*? Or did it develop only recently, and rapidly, perhaps with the appearance of modern humans not much more than 100,000 years ago, or even later? Because language, unlike, say, technological skill, is ephemeral in the prehistorical record, anthropologists have to seek clues in indirect evidence, in two realms. The first is in our ancestors' anatomy, such as brain size and configuration and the vocal tract. The second is in aspects of behavior that might reflect language capabilities, including culture and artistic expression.

Taken as a whole, our ancestors' anatomy points to the slow, gradual evolutionary scenario. For instance, the size of the brain increases above the australopithecine level with *Homo habilis,* measuring 660 cc at around 2 million years ago; early *Homo erectus* is around 850 cc; late *erectus,* at 400,000–100,000 years ago, has risen to 1100 cc; and *Homo sapiens* is 1300 cc. If brain size reflected language capacity in a direct, one-to-one manner, the conclusion is that the Mojokerto child would have been developing vocal skills that were in some sense verbal and representational, not just the limited vocalization of other large primates.

Moreover, the configuration of the brain's lobes took on a distinctly human aspect in earliest *Homo,* with prominent frontal lobes and relatively small occipital lobes (at the back of the brain), as compared with a more apelike aspect that is found in australopithecines, where the occipital lobe is relatively bigger than in humans and the frontal lobes relatively smaller. This is not necessarily indicative of an incipient language capacity, but it is at least suggestive. The left hemisphere of the brain of early *Homo* was slightly larger than the right, as is the case in most modern people (right-handers), at least partly because the left brain is where important language functions are located. Most language-associated mental machinery is buried within different parts of the brain and is invisible to paleoanthropologists, who can see only the overall shape of fossil brains. However, a lump on the side of the brain and toward the front, known as Broca's area, is associated with some aspects of language function. It is not the clear signal of language abilities that anatomists once thought it was, but again it is suggestive. Broca's area is visible in 1470, that is, *Homo habilis,* and in subsequent *Homo erectus* skulls that are complete enough that the lump can be looked for.

Brain size and form therefore support the notion that language arose gradually, beginning early, and so too does the anatomy of the larynx, the voice box. Although the tissues of the voice box don't fossilize, because they are constructed from soft tissues such as cartilage, the underside of the cranium is indicative of the *position* of the larynx in the throat. This is important because in all mammals apart from humans, the larynx is high in the throat. The low position in humans permits a great range of sound production, which is obviously important for language. (It also uniquely confers on humans the liability of choking while trying to swallow food and breathing simultaneously.) In species with a high larynx, the underside of the skull is relatively flat, whereas in humans it is flexed, like a bridge. Anthropologists such as Philip Lieberman, Edward Crelin, and Jeffrey Laitman have examined the contours of dozens of fossil human skulls, ranging from *Homo habilis* to Neanderthals, and have found that the degree of flexing becomes gradually more pronounced through time, beginning with *Homo erectus.*[8] A signal of gradual development of language abilities? Again, it is suggestive.

There is, however, one anatomical signal that some anthropologists interpret to mean that the Mojokerto child, and all *Homo erectus* individuals, were no better than apes in their vocalization. Once again the source is the Turkana boy's skeleton.

The beautifully preserved vertebrae of the thorax region of his body reveal that the hole through which 1470's spinal cord ran was about half the size that it is in modern humans. Ann MacLarnon, of Whitefields College, London, who studied the boy's vertebral column, points out that in modern humans the hole in the spinal cord in the thoracic region is enlarged to accommodate the many nerves that control muscles in the chest wall and the abdomen. These muscles are important in the fine orchestration of breathing necessary for producing complex, spoken language. Given the smaller hole, MacLarnon speculates that *Homo erectus* did not have a well-developed capacity for spoken language.

The anatomical evidence has therefore spoken mostly in favor of the slow, gradual scenario. What of archeological information, insights into behaviors that might be indicative of language capacity? Here, the weight of evidence is reversed, with a recent, rapid origin being favored.

Our ancestors made stone tools for increasing their control over a practical realm of their lives, principally access to foods of various kinds. The complexity of tool assemblages increased over time, from the very simplest pebble tools of the Oldowan to the beautifully detailed implements of the Upper Paleolithic period, which began 40,000 years ago. If, as some archeologists have argued, the increasing complexity reflects not only greater technical skills and demands but also increasingly stringent social norms, then tools would speak of language, too. What is the trajectory of change in complexity? Slow for 2 million years, and then increasing acceleration beginning 200,000 years ago, and going off the charts in the Upper Paleolithic. Styles changed rapidly in this last period, and there were regional fashions, redolent of strongly expressed culture—expressed, no doubt, in rich, fully modern language. If this trajectory of increased technological skill directly reflects the path of language evolution, we would have to say that the loom of language is a recent acquisition in the human mind, and that the Mojokerto child was more like an ape than like a human in her vocalizations.

The same is true for artistic expression. The symbolic abstraction we associate with artistic expression is unthinkable—literally as well as figuratively—in the absence of language. The two must have evolved in concert, say the archeologist Iain Davidson and the psychologist William Noble of the University of New England, Australia. Symbol making, in terms of engraved or painted images, embodied—and required—the abstraction that is central to human language.[9] Such tangible symbol making appears relatively suddenly, and unequivocally, in the Upper Paleolithic in Europe, beginning just 40,000 years ago. This means, argue Davidson and Noble, that language first appeared this late in prehistory, too.

Randall White, an archeologist at New York University, agrees. He looks at aspects of human behavior as revealed in the archeological record and concludes that earlier than 100,000 years ago, there was "a total absence of anything that modern humans would recognize as language."[10] White lists seven areas of archeological evidence that, in his view, point to dramatic enhancement of language abilities coincident with the Upper Paleolithic, perhaps with a slow fuse having been lit some 65,000 years earlier. The first is evidence of deliberate burial of the dead, which begins almost certainly in Neanderthal times but becomes refined (with the inclusion of grave goods) only in the Upper Paleolithic. Second, artistic expression—image making and bodily adornment—begins only with the Upper Paleolithic (and Later Stone Age, in Africa). Third, he cites the sudden acceleration in the pace of technological innovation and cultural change in the Upper Paleolithic. Fourth is the evidence for regional differences in culture, which he says is an expression and product of social boundaries. Fifth, evidence of long-distance contacts—exotic objects, such as rare stones, traded between groups—becomes strong. Sixth, living sites significantly increase in size (complex language is a prerequisite for planning and coordination). And seventh, technology moves from the predominant use of stone to include other raw materials, such as bone, antler, and clay.

It is therefore not surprising that most archeologists support the recent, rapid scenario of language origins. What does this mean in relation to the anatomical evidence, which almost as strongly favors the early, gradual scenario? One thing that it most obviously means is that at least some of the lines of evidence that have been taken as indicative of language capacity must be indicative of something else instead. The challenge is to discover which features of fossils or aspects of behavior truly reflect language. Although the archeological evidence looks persuasive of a rapid, recent event, it may be that the event is at least in part cultural, not entirely evolutionary. By analogy, we could look at the degree of technology apparent in society from the Middle Ages to today. The pattern of change is of a long period of little change and then an escalating trajectory of innovation, beginning with the Industrial Revolution. And yet people are not inherently smarter now than they were half a millennium ago. The pattern reflects cumulative technological knowledge that, having passed a certain threshold, feeds on itself in an ever more voracious feedback loop.

No one would suggest that the pattern of change in technology or art that is seen in the prehistoric record is simply cultural. There is overwhelming evidence of biological change through that period, of course. But it is prudent to at least acknowledge that something equivalent was going on in our history, and that these strong patterns of technological and artistic behavior may be blinding us to more subtle signals. Not so subtly, Dean Falk, an expert on fossil brains, asks, "If [early humans] weren't using and refining language I would like to know what they *were* doing with their autocatalytically increasing brains."[11]

Until it is proven otherwise, it is reasonable to argue that the initial trigger to increasing brain size was technological, the need to manufacture and use tools. Very soon, as the social and economic milieu of a primitive hunting and gathering way of life developed, there would be an evolutionary advantage in those individuals who could build ever better mental models of themselves, their worlds, and the worlds of others with whom they interacted. Natural selection would encourage the evolution of language and consciousness, beginning in rudimentary form and becoming more advanced. In this case, Haeckel was wrong when he called his "missing link" *Pithecanthropus alalus,* the mute ape-man, because there is a strong possibility that she did have some command of language.

There may well have been a threshold crossed quite recently, where the cognitive components associated with language and consciousness that had been emerging gradually but steadily suddenly coalesced, permitting something that was far more complex, far richer, than had been possible previously. Such a coalescence may have been the force behind the recent origin of fully modern humans, people like us.

II

The Origin of Modern Humans

If anthropologists' passion for pursuing the roots of language is set at a high temperature, their ardor for wrestling with the origin of modern humans—that is, *Homo sapiens*—must be described as being off the top of the scale. During the 1980s and 1990s there was a blizzard of research papers, conferences, and academic books on the topic, characterized, it has to be said, as much by contention as by measured debate, and it continues still. And if the volume of coverage in the popular media is any guide, there is a deep and unquenched thirst among a more general audience for learning about how we finally came to be what we are today. At the very least, we can say that the topic obviously holds some kind of wider fascination. One reason for public interest in this particular realm of academia is, no doubt, the unusually combative tenor of the debate, a clear sign of a roiling controversy. Controversies make good copy and catch the interest of nonscientists to a degree that dispassionate exposition of ideas seldom does, unfortunate though this may be. It is also true that most people—academic and otherwise—are genuinely interested in our origins: such curiosity is a clear facet of our nature as humans.

But the fact that the academic debate over modern human origins has been more like a bar brawl than objective scientific discourse says something about the nature of the topic itself, and what supposedly objective scholars bring to it. How, otherwise, could the same evidence be scrutinized with the same techniques of analysis by two individuals who then reach entirely opposite conclusions? Science progresses by the constant collision of different opinions, of course, with differences eventually being resolved. But the fact that the topic at hand here—ourselves—is so personal, if in a collective way, means there is more than a little garnish of subjectivity on this supposedly objective scientific inquiry.

Unlike the question of the roots of language, which struggles to find relevant evidence in the prehistoric record, the question of how *Homo sapiens* arrived in the world is addressed by a comparatively rich collection of clues—if only those clues could be deciphered correctly. There are fossils from three continents, and stone tools, and—the most recently appreciated sources of answers to questions about ancient events—the genes of modern people. Within this confection of evidence, the Mojokerto and Sangiran fossils, for which we produced such surprisingly ancient dates, play a major—some have said decisive—role in resolving the issue.

Anthropologists agree on the overall anatomical and behavioral shifts that accompanied the evolutionary transformation from *Homo erectus* to *Homo sapiens.* The skeleton of *Homo sapiens* is much less robust than that of *Homo erectus,* meaning that the bones are not as thick and the joints not as massive, which indicates that we are a lot less physically strong than our forebears. The skull housed a bigger brain and was more vaulted in shape, rather than being long and low; and the brow ridges—so characteristic of *Homo erectus*—disappeared. In the behavioral realm, *Homo sapiens* crafted finer and more diverse tool technologies, using many different types of material, not just stone; they pursued more efficient foraging strategies; social organization was more complex; language became fully developed; and artistic (symbolic) expression exploded. There is also agreement that beginning about 400,000 years ago and continuing until some 100,000 years ago, new human forms, known as archaic *sapiens* and combining aspects of ancient and modern anatomy, appeared in Africa and Eurasia. The disagreement—deep intellectual chasm, in reality—is over this evolutionary transformation: when it took place, where it took place, how it took place, and where archaic *sapiens* species fit into the picture. Tightly entwined in this issue is the Neanderthals (a type of archaic *sapiens*) and their fate: Were they ancestors of modern Europeans? Or did they go extinct without issue?

The debate over the origin of modern humans has a long history, beginning, appropriately enough, with the discovery of the Neanderthals. Theories have ebbed and flowed, with the inferred fate of the Neanderthals fluctuating back and forth many times between ancestor and extinct side branch. In the modern debate, there are two major, opposing hypotheses, and if a show of hands were to be called for among anthropologists, most would declare that the Neanderthals were an evolutionary dead end, ancestral to nothing.

One hypothesis argues that the transformation occurred as a gradual change within all populations of *Homo erectus* wherever they existed, leading to the near-simultaneous appearance of multiple populations of modern humans in Africa and Eurasia. In this view, the genetic roots of modern geographical populations of *Homo sapiens* are deep, reaching back to the earliest populations of *Homo erectus* as they became established throughout much of the Old World. This is known as the Multiregional Evolution Model. The competing hypothesis views modern humans as having a recent, single origin (probably in Africa), followed by a population expansion that replaced established nonmodern populations in the rest of the Old World. In this scenario, the genetic roots of modern geographical populations of *Homo sapiens* are very shallow, going back perhaps 100,000 years. This is called the Single-Origin, or Out of Africa, Model. There are other hypotheses, which combine elements of these two positions to different degrees, but the multiregional and single-origin models most clearly illustrate the lay of the intellectual landscape.

The multiregional hypothesis was the first comprehensive theory of the origin of modern humans, and its history goes back more than a century, to when Gustav Schwalbe, the Strasbourg anatomist, argued that there was an evolutionary sequence beginning with *Pithecanthropus,* leading through Neanderthals, and continuing on to modern humans. Fifty years ago the German anatomist Franz Weidenreich established more firmly this so-called unilinear view. He envisaged parallel evolutionary lineages in various regions of the Old World leading through separate archaic *sapiens* forms to the geographical variants of modern humans. Weidenreich's proposal came to be known as the Candelabra model of modern human origins: drawn schematically, the long regional ancestries look like an array of candles.

Weidenreich was aware that by suggesting that each modern geographical population traces its origins back through Neanderthal-like and *Homo erectus* precursors, he might be understood as saying that modern races have separate origins, even that they are separate species. In 1949 he explicitly stated that this was not his view. Nevertheless, in 1962 the University of Pennsylvania anthropologist Carleton Coon came close to proposing what Weidenreich had warned against. Coon argued not only that racial differences were ancient but also that some races had achieved sapienshood earlier than others. The notion that extant racial groups have been genetically separate for at least a million years and that some were relatively recently evolved lent itself readily to the inference of deep differences between the races.

Despite Weidenreich's efforts, the so-called unilinear point of view was slow to take hold. Meanwhile, a confluence of events through the 1940s, '50s and '60s led to the development of several other hypotheses. One included the proposition of the evolution of modern forms earlier than Neanderthal, which therefore left Neanderthals as an extinct side branch; another suggested that the "classic Neanderthals" of western Europe were the evolutionary product of less extreme forms in eastern Europe and the Middle East, which also gave rise to modern humans.

The emergence of the unilinear hypothesis as the most prominent among its competitors came about in the 1960s, principally through the efforts of Loring Brace of the University of Michigan. In a 1964 paper, "The Fate of the 'Classic' Neanderthals,"[1] Brace argued that Neanderthal anatomy had been mistakenly interpreted as extreme, and that it could be seen as ancestral to that of later European modern people. By the late 1960s, therefore, the Neanderthals had been restored to—in many people's eyes—their rightful place: as direct ancestors of modern humans. Fossils that had been discovered in Europe, Africa, and Asia during the first half of the century were now interpreted by Brace and his supporters within the unilinear, or single-species, theory as evidence of evolution toward *Homo sapiens* in many different parts of the Old World, as we noted in chapter 8.

The single-species hypothesis was put to rest in 1975, when

new fossil evidence showed unequivocally that more than one species of human had lived side by side in Kenya some 1.8 million years ago. Milford Wolpoff of the University of Michigan, a co-proponent with Brace of the single-species hypothesis, nevertheless insists that the hypothesis holds for the later stages of human prehistory: *Homo erectus* populations were engaged in a slow transformation into *Homo sapiens* throughout the Old World, he says. The hypothesis suggests that the anatomical variation we see in modern populations evolved in the different geographical locations long, long ago and has been maintained. This purported anatomical persistence is known as regional continuity.

In updating Weidenreich's model, Wolpoff, joined by Alan Thorne of Australian National University, Canberra, suggests that the different geographical populations were not entirely isolated from one another, and that limited gene flow—that is, intermingling of people—between populations has been important. The *erectus*-to-*sapiens* transformation was therefore a balance between the maintenance of distinctive regional anatomical features through partial population isolation persisting for very long periods of time, and the maintenance of a genetically coherent network of populations in the Old World through significant interbreeding.

The multiregionalists argue that the evolutionary transformation from *Homo erectus* to *Homo sapiens* involved continuous change within a genetically coherent lineage, even though it was geographically dispersed. This being the case, there is no clear break between *Homo erectus* and *Homo sapiens,* merely an evolutionary continuum. The inescapable consequence of this, say the multiregionalists, is that there is no valid reason to distinguish between species. All forms of humans, from 2 million years onward, were members of a single species, *Homo sapiens*. Not surprisingly, many people balk at the suggestion that people living today and those who lived in, for instance, Java almost 2 million years ago are members of the same species. Such a suggestion mocks biological reality.

The Out of Africa hypothesis is the younger of the two models, in terms of history. It goes back to the ideas of Louis Leakey in the 1960s, who preferred to believe that certain *Homo erectus*-like humans in Africa were more likely ancestors for modern humans than the *Homo erectus* fossils of Asia; the latter, he said, were an evolutionary dead end. In its modern incarnation, the Out of Africa hypothesis says that anatomically modern humans evolved in a small geographical region, probably in Africa, and that descendants of these founding members of *Homo sapiens* moved into the rest of the Old World quite rapidly, substantially replacing the existing archaic *sapiens* populations. There might have been some interbreeding between *Homo sapiens* newcomers and incumbent archaics, says Christopher Stringer of the Natural History Museum, London, the most prominent proponent of the hypothesis, but its long-term effects were minor. In contrast with the multiregional evolution model, the Out of Africa hypothesis views geographical variation among modern populations as having developed very recently, certainly within the last 100,000 years and perhaps closer to 10,000 years ago.

How to choose between the opposing models? One has to look at the predictions they make for what should be seen in the prehistoric record, and check how the predictions match reality. Scientific hypotheses live and die by their predictions of the real world. For instance, if modern humans evolved the way the Out of Africa hypothesis suggests, the record should look as follows: Anatomically modern humans should appear in one geographical region—probably Africa—earlier than in others; forms that are transitional between archaic and early modern anatomy should be found only in Africa; there should be no necessary link between anatomical variants in modern populations and ancient populations; there should be little or no evidence of hybridization between archaic and early modern populations. Not surprisingly, predictions of the multiregional evolution model are more or less directly opposed to these: Anatomically modern people will appear more or less simultaneously throughout the Old World; ancient and modern species should not overlap in time—that is, *Homo erectus* and *Homo sapiens* should not coexist; transitional forms should be found in all regions of the Old World; and there should be strong signals of regional continuity of characteristic anatomy.

We will not spill more ink than is necessary on a topic on which enough ink has been spilled already, that is, in determining how the hypotheses measure up to reality. The short answer is, opinions differ. We can say that by the mid-1980s, the battle between the two models was vigorous but not out of hand. The Out of Africa model was, however, gaining ascendancy. By most anthropologists' assessment, for instance, the earliest known anatomically modern humans are from Africa—in South Africa and Ethiopia, both dating to about 130,000 years. In most geographical regions there is no good evidence of regional continuity, though multiregionalists would dispute that. Evidence of hybridization is slim at best. And archaic *sapiens* forms outside of Africa could just as plausibly be argued as being separate species that became extinct, rather than the true transitional forms that multiregionalists believe.

Just when it was beginning to seem as if the same evidence—and the same opposing opinions—must continue on forever, a new line of inquiry entered the scene. In January 1987, Allan Wilson and two colleagues from the University of California, Berkeley, published a paper in the journal *Nature,* titled "Mitochondrial DNA and Human Evolution." Its effect was to crank up the decibels, and provoke chromatic vocabulary, in the debate. In the paper, Wilson and his colleagues wrote: "[The] transformation of archaic to anatomically modern forms of *Homo sapiens* occurred first in Africa, about 100,000 to 140,000 years ago, and . . . all present-day humans are descended from that African population."[2] The Berkeley group saw no evidence of ancient mitochondrial DNA (that is, from *Homo erectus,* for instance) in modern populations. This implies that there had been complete replacement of existing archaic populations by incoming modern humans, and no hybridization.

The Wilson lab's results offered strong support for the Out of Africa hypothesis and none for the multiregional evolution hypothesis. And it initiated a scientific saga that is as much sociology as it is science, and illustrates why humans, no matter how scientific they are, often find it hard to be objective about this oh-so-important aspect of themselves, their origins.

The Rise and Fall of Mitochondrial Eve

Mitochondria, the energy-producing organelles present in all cells in the body, contain circular molecules of DNA. Mitochondrial DNA is useful in reconstructing recent evolutionary histories of modern populations, for two reasons. First, it is inherited strictly through females, which makes reconstruction of evolutionary patterns more straightforward than with nuclear DNA, which recombines from both parents at each generation, thus scrambling the DNA to some degree. Second, mitochondrial DNA accumulates mutations at about ten times the rate of nuclear DNA, and therefore offers a "fast-ticking" molecular clock that can record relatively recent events. In terms of evolutionary time, events within the last several hundred thousand years are relatively recent.

By analyzing about 9 percent of the DNA sequence in 147 individuals from around the world (using a technique known as restriction fragment length polymorphism), Wilson and his colleagues found that, first, all the DNA types were young, with no ancient ones evident; second, linking the types of DNA in a genealogical tree produced two groups, one containing representatives of all populations, and a second containing only Africans; and third, the amount of genetic variation in the DNA types was greater in Africans than in any other population. This was very much what the predictions of the Out of Africa hypothesis say, but cast in genetic terms: that is, modern humans evolved recently (indicated by the recent age of the DNA) and in Africa (because this is where the most variation is today, indicating that this population existed first).

Wilson's proposition came to be called the Mitochondrial Eve hypothesis, for reasons that have to do with a mixture of science and our love of telling stories. Tracing the genealogical tree of mitochondrial DNA types effectively leads back to the DNA of a single female who lived many generations ago. This is best explained by analogy. Imagine a population of, say, 5,000 mating couples, each with a different family name. Now imagine that as time passes the population remains stable (each couple produces only two offspring). In each generation, on average, one quarter of the couples will have two boys, one half will have a boy and a girl, and one quarter two girls. In the first generation, therefore, when the children marry, one quarter of the family names will be lost, from the families that had only girls. As each generation passes, more losses will occur, but at a slower rate. (After all, more of the couples in each successive generation share the same name, so the name is not eliminated entirely when one of the couples has two girls.) After about 10,000 generations (equal to twice the number of original mothers), only one name will remain. The same pattern holds for the loss of mitochondrial DNA types, except that the transmission is through the female line.

For this reason, Wilson often described the single female from whom we all derive our mitochondrial DNA in the modern world as "one lucky mother," because the loss of DNA is pure chance.

Wilson was very aware that his one lucky mother way back in prehistory was in the company of many other mothers, whose mitochondrial DNA got lost on the evolutionary journey. She was not Eve alone with her Adam. But, unfortunately, the language and imagery of an Eve was too strong to be ignored, and it introduced a confusion that was not confined to nonscientists. A year before the publication of the *Nature* paper, a reporter for the *San Francisco Chronicle* wrote a front-page story about the work, titled "The Mother of Us All—A Scientist's Story." Charles Petit, the reporter, described the Berkeley team's conclusions, and noted that "the dramatic, controversial claim of a fairly recent African Eve as the very very great-grandmother of all humans is sure to stir up an old debate in paleoanthropology." The correct technical term for that single female is the "mitochondrial DNA lineage coalescence point," but Petit used the catchier term "African Eve" instead, knowing that the drier alternative would probably not get past the editor—and in any case Eve's is such a primal story in our culture that it was impossible to pass up the allusion. Even scientists got caught up in this imagery and incorporated it in the way they thought about the problem. Invoking the name "Eve" also led writers to describe her in ways that the evidence did not imply, as we'll see below.

When a population falls to low numbers—a few hundred or a few thousand—it is described as a population bottleneck. The idea of an extreme population bottleneck therefore became identified with the Mitochondrial Eve hypothesis—after all, Adam and Eve represent the most extreme bottleneck of all. No matter how many times Wilson said that Eve had been just one of perhaps 10,000 females in the population at the time, the Eve imagery prevailed.

Critics of the Eve hypothesis were quick to say that the hypothesis must be wrong because there is no genetic evidence for an extreme bottleneck in human prehistory. For instance, the geneticist Francisco Ayala of the University of California, Irvine, wrote a paper in the journal *Science* entitled "The Myth of Eve." Although he accepted the claim that Africa was probably the birthplace of modern humans, he went on to say that "the weight of the evidence is against a population bottleneck before their emergence."[3] The model was being condemned, not for something it predicted as a scientific hypothesis, but for something inherent in its popular sobriquet.

Eve quickly caught the media's imagination and even received the ultimate public accolade, that of being the subject of a cartoon in the *New Yorker.* Eve also made a cover story in *Newsweek,* a year after the work was published in the scientific literature. The text acknowledged that Eve was not the only female living back 150,000 years ago, but it suggested that she "was the most fruitful" and that she left behind "resilient genes" that are carried by all of humankind.[4] In retrospect it might appear that way, but in fact, as explained in the analogy above, the survival of the mitochondrial DNA lineage has to do with pure chance, not greater fertility or superior genes. But we'd expect Eve, mother of us all, to be fruitful and strong, wouldn't we?

Not surprisingly, proponents of multiregionalism were not happy with the Mitochondrial Eve hypothesis, not least because, just as had happened two decades earlier with the origin of the human family, here was a bunch of geneticists coming along and telling anthropologists that they had got the story wrong. And once again Allan Wilson was in the thick of it. A passage from an article in *Discover* in 1990, written satirically, was meant to explain anthropologists' negative reaction to Eve:

> Her name was Eve, and she was trouble from the start. Sure, the public loved her: she was brash, sexy, and surprising, with a body of data you could reach out and grasp and implications that just wouldn't quit. She made the cover of *Newsweek.* She even got on Johnny Carson. Not too shabby for a human-origins hypothesis born in a biochemist's beaker. But for paleoanthropologists, the hard-boiled types who earn their living making dead men talk, Eve spelled poison forward and backward. She was an interloper, a biochemical bauble, a dolled-up set of assumptions masquerading as a breakthrough. To them, the only good Eve was a refuted Eve.[5]

That hoped-for refutation was what appeared to have happened at a symposium of the American Association for the Advancement of Science in New Orleans in February 1990. The collective message of the group of anthropologists who ran the session was twofold: first, that fossils, not molecules, were the only evidence that would settle the matter at hand; and second, that the fossil evidence "proves that the Eve hypothesis *must* be wrong," as they concluded in a collective statement handed to the press. The symposium was widely reported, and a quote from the London *Times* gives a taste of the tenor of those reports: "A group of specialists on fossils said that studies of skulls and other remains of humanlike creatures in Asia and Europe showed that the Garden of Eden theory must be wrong."

The anthropologists had prepared careful summaries of their arguments and handed them out as a single document to the journalists at the meeting. The document was entitled "Eve: The Fossils Say No." What the document didn't explain was that every one of the speakers was a proponent of the multiregional evolution hypothesis and so had a vested interest in proving Eve wrong. A different group of anthropologists, this one selected by, say, Christopher Stringer, would have reached a different conclusion: "Eve: The Fossils Say Yes."

Milford Wolpoff was the symposium's organizer, and later he told a reporter that the event was "a sales pitch," not the objective assessment of all the fossil evidence that the program had implied. "We planned the whole thing, rehearsed it, worked over the exact phrasing," he explained. "We felt we had to do this, because we were becoming victims of the complexity of our ideas . . . We think the fossils absolutely support our position."[6]

Wolpoff used some evocative—or, more accurately, provocative—language when he compared his own model to the Mitochondrial Eve and the Out of Africa hypotheses, which were by then closely identified with each other. "Our hypothesis of regional continuity may not be as sensational as this idea of killer Africans sweeping out across Europe and Asia, overrunning everybody," he said. "There is no way one human population could replace everybody else except through violence. The people advocating replacement have to come to terms with what they are saying. I'm just glad it's them and not me."[7] Invoking images of genocide, similar perhaps to the near genocide of native peoples in the Americas and Australia in recent historical times, may make good copy, but scientifically it isn't true. For instance, Ezra Zubrow, of the State University of New York, Buffalo, has developed demographic models that show that even a very small difference in the efficiency of exploiting food resources between two populations can result in the rapid extinction—within a single millennium—of the less efficient of the two. The rate of extinction is consistent with what is seen in the fossil record.[8]

An Obituary, Prematurely Penned

For a period of about three years, Wilson and his hypothesis basked in a glow of approval—not, as we just saw, from the multiregionalists, but in the more general domain of spectators of the debate. Then it all began to fall apart. The study was criticized for not sampling native African populations (he had used African Americans), and, worse, it was claimed that the data had been incorrectly analyzed so that the conclusions weren't statistically significant. Just as quickly as Mitochondrial Eve had been embraced by the press (albeit for the wrong reasons), she was denounced. "Critics batter proof of an ancient African Eve," ran a headline in a *New York Times* article. The text said, "The serpent of uncertainty has slithered into the garden. Its bite has undermined a critical statistical foundation for the Eve hypothesis."[9] The author, John Noble Wilford, one of the best informed and most thoughtful of newspaper science writers, could not resist developing the Garden of Eden imagery. Continuing the biblical imagery, an article in the *Chronicle of Higher Education* stated that "Black Eve had fallen from grace."[10]

Wilson and his team had brought condemnation on themselves because of the way they had dealt with the huge amount of data flowing from the analysis of mitochondrial DNA in the context of seeking to uncover evolutionary history. This involves looking at the genetic variation among modern populations and trying to build trees with the fewest number of genetic changes that link the genetic variants. There are millions of possible trees, so many that it was impossible for Wilson and his colleagues to test them all with the available computational power. The shortcuts they took unfortunately made an extremely complex picture look simple. Wilson died at about this time, leaving his colleagues, particularly Mark Stoneking, now at Pennsylvania State University, to carry on the work.

Stoneking and his new colleagues were prompted to reexamine their analyses, and soon conceded in a short item in *Science* in February 1992 that the conclusions about an African Eve could no longer be said to be statistically significant. Nevertheless, Stoneking insisted, the overall conclusion was still likely to be correct. Wolpoff didn't agree, and said so unequivocally at a conference not long after Stoneking's *Science* paper: "It's over for Eve."[11]

Eve's obituary was, however, premature. True, this setback made geneticists realize that genetic data other than those from mitochondrial DNA would be necessary to obtain clearer answers. And this is precisely what began to happen, with laboratories around the world scrutinizing dozens of genes for further clues to our recent evolutionary history. This included looking at the male, Y, chromosome, which is the male equivalent of mitochondrial DNA in that it gives information about the male lineage. Inevitably, this was cast as the search for Adam. The result, captured in the title of an article in *Science,* gave comfort to Eve: "Y Chromosome Shows That Adam Was an African."[12]

And there was to be more comfort from a comparison of the genetic information of all the genes so far analyzed. Using a different analytical perspective, Maryellen Ruvollo pointed out that, in terms of genetics, the key difference between the multiregional and single-origin hypotheses was one of timing. If the multiregional evolution hypothesis was correct, then encrypted in our genes would be a signal of that earliest migration of *Homo erectus* out of Africa some 2 million years ago. In other words, there would be a signal of an ancient event. By contrast, if the single-origin, or Out of Africa, hypothesis was correct, the signal would be of a recent event, the rapid replacement of existing archaic populations. Ruvollo's conclusion: "The rapid replacement hypothesis is the likeliest model for modern human origins."[13] Mitochondrial Eve might not have been able to produce an unequivocal answer all by herself, but, through her flaws, she prompted a little help from others, including Adam. From a geneticist's point of view, anyway, the debate is leaning very heavily in one direction—not unequivocally, but "likeliest."

The kicker in this story has something of a surreal edge to it, and it comes, appropriately enough, from the Neanderthal individual whose bones lay entombed for 60,000 years in that little cave in the Neander Valley. For more than a century, anthropologists have pored over these ancient relics, visually scrutinizing every nuance of their form, applying calipers for precise measuring, seeking answers: Do modern Europeans carry some of the same genes that, long ago, gave life to this Neanderthal Man? Are modern Europeans descendants of the Neanderthals, as the multiregional evolution hypothesis argues? In a marriage of high tech and high drama, researchers at the University of Munich cut away a small piece of the Neanderthal's right upper-arm bone and went in search of Neanderthal genes.

Biologists have watched with growing fascination in recent years as some of their colleagues have developed methods for extracting genetic material—small fragments of genes—from organisms long dead. Known as ancient DNA research, the new science was the inspiration for Michael Crichton's book *Jurassic Park.* Although not quite as dramatic as that wild fantasy, the Munich researchers, led by the ancient DNA research pioneer Svante Pääbo, extracted a small amount of DNA from the Neanderthal bone and compared it with modern genes. Scientists being able to do genetics on ancient members of the human family was a huge first in anthropology. "It's one of my dreams that this would be possible," said Chris Stringer when he heard of the achievement. "For human evolution, this is as exciting as the Mars landing."[14]

Equally exciting were the results, not least because they were quite clear, even unequivocal: the Neanderthal DNA was substantially different from modern DNA, sufficiently so as to rule out all but the slimmest of possibilities that Neanderthals were our ancestors. In the scientific paper in which they reported the study, Pääbo and his colleagues wrote that "this suggests that Neanderthals went extinct without contributing mitochondrial DNA to modern humans."[15] The eminent Stanford University geneticist Luca Cavalli-Sforza, whose life's work has been the study of genetic patterns in human populations, was so impressed with the result that he told an editor at *Science* magazine that the work "destroys one of the fortresses of the regional continuity model."[16]

Proponents of the Out of Africa model (based on fossil evidence) can therefore gain comfort from the messages flowing from genetics labs this past decade and a half, while multiregionalists are left to explain the apparent lack of support for their hypothesis.

A Change of Date, a Change of Heart

While supporters of the Out of Africa hypothesis were watching the Mitochondrial Eve saga unfold, with its heartening messages for their own position, events in the anthropological realm were smiling on them, too. When you change the dates applied to certain fossils, the entire evolutionary picture can change, of course. So it was with the question of the origin of anatomically modern humans, in two places: in the Middle East and in Java. Supporters of multiregionalism for a long time held these two geographical regions to be powerful pillars supporting their claims. In the first, they suggested, was evidence of a transition from Neanderthals to modern humans. The second was said to show a clear continuity of anatomical features from *Homo erectus* in Java through to the earliest aborigines of Australia.

Israel has long been a favored locale for anthropologists, but usually with biblical-age relics in view. But the country is also rich in sites—principally caves—containing recent human ancestors, several of which apparently were deliberate burials. In the 1930s, two caves on Mount Carmel, Skhul and Tabun, yielded partial human skeletons, those in Skhul being distinctly anatomically modern and those in Tabun being Neanderthals. Then, between 1935 and 1975, excavations at Jebel Qafzeh, near Nazareth, produced an even bigger cache of modern forms, while more Neanderthals turned up in the Kebara Cave, also on Mount Carmel, and at Amud, near the Sea of Galilee. It wasn't possible to put certain dates on these specimens initially, but generally it was estimated that the Neanderthals were around 60,000 years old and the modern forms close to 40,000, a perfect time progression from Neanderthals to moderns, as posited by multiregionalism.

As new methods for dating developed in the 1980s—principally electron spin resonance and thermoluminescence—it became possible to release these fossils from a virtual time limbo, with dramatic effect. It turned out that the modern human individuals of Skhul and Qafzeh all lived about 100,000 years ago, while the Neanderthals of Tabun were a little older. An ancestor/descendant relationship between Neanderthals and moderns was therefore still feasible. The Amud and Kebara Neanderthals, however, were found to be at most 60,000 years old. With modern forms *predating* some of the Neanderthals, it is impossible for Neanderthals to be ancestral to modern humans in this region. Multiregionalists say that this is not an issue, because they make no distinction between the individuals: all are members of an anatomically variable population, they claim. But that would require a degree of variability unknown in any other human population, and so the argument looks like special pleading. If, as seems highly probable, the Neanderthals did become extinct without issue, then they should be recognized as a distinct species, *Homo neanderthalensis,* just as the Irish anatomist William King proposed way back in 1864.

One interesting aside here is that although there presumably were significant biological differences between Neanderthals and modern humans in this region, they both apparently made and used the same kind of tool technology, a form that Neanderthals had been using since their earliest times, at least 150,000 years ago. This disjunct between modern anatomy and modern behavior is one of the puzzles that attend modern human origins, and it is not just in the Middle East. Although the modern human form apparently evolved as much as 130,000 years ago in Africa, the characteristically modern human toolkit—with fine blades, and bone and ivory implements—didn't really become established until at least 60,000 years later. There is a caveat, however, in that archeologists' views might be skewed by the paucity of well-preserved archeological sites in this time period. And, little by little, glimpses of modern human tool-making behavior are turning up at several sites in Africa, close to 100,000 years old, and one of them considerably older than that.[17] Perhaps this discontinuity is more apparent than real.

Java holds a special place in multiregionalists' thinking. "In each region of the world, we have uncovered links that tie living populations to their local antecedents," notes Milford Wolpoff and Alan Thorne. "The most convincing evidence comes from Asia."[18] Wolpoff and Thorne speak of a series of fossils linked by a common anatomy, from the earliest to the most recent times, from *Homo erectus* of Sangiran, which, until our recent redating, was thought to be about 700,000 years old; through the collections of skulls at Ngandong, until recently thought to be between 400,000 and 100,000 years old; and finishing with the Australians of 10,000 years ago. "Early Indonesians have large projecting faces, with massive rounded cheek bones and large teeth," observe Wolpoff and Thorne. "The face bears a number of small but important features: a 'rolled edge' on the lower margin of the eye socket, a distinctive ridge on the cheek bone and a nasal floor that 'flows out' smoothly onto the face. These and other traits combine to create a special Indonesian variation on the *Homo erectus* theme."[19]

The complex of features seen in the crania of early Indonesians continues on through the Ngandong people and into recent Australians, say Wolpoff and Thorne, leading them to describe it as the "mark of ancient Java," a phrase coined by the Australian anthropologist N. W. G. Macintosh in 1965. Critics of this scheme have repeatedly pointed out that this putative "continuous sequence" is more gap than substance: three single points spread over almost a million years. This is a very static view of a population that surely experienced a degree of flux over that extremely long period of time. With our new dating of the Sangiran fossils, the problem becomes ever more acute, because we are now talking about a time period approaching 2 million years. "Can anyone seriously propose that the lineage of Australian aborigines could go back that far?" Christopher Stringer asks incredulously.[20]

The reason for the incredulity is simple. The multiregional hypothesis demands that there be genetic continuity, or gene flow, throughout all Old World populations of *Homo erectus,* so that evolutionary change in one region ripples into other regions. And this must persist over a very long period of time. Many geneticists question whether such continuity is possible at all, in a population spread across three continents, let alone whether it can be maintained for a million years, as the previous scenario held. Maintaining that continuity for twice that length of time is simply off most population geneticists' radar screens of possibility.

Wolpoff dismisses the notion that the new Java dates are a fatal blow, but concedes that they "affect it in an important way." He suggests that anthropologists should think of multiple dispersals out of Africa, not all of them successful. "The first successful expansions were made possible by the development of Acheulean industry," he argues, appealing to the technological sophistication hypothesis mentioned earlier. "It gave the people a competitive edge over populations that were established from earlier expansions . . . Yes, these earlier-established populations might have been replaced by the Acheulean people," concedes Wolpoff. "The Acheuleans in eastern Asia didn't make hand axes because they used bamboo instead."

There is a wry irony in hearing Wolpoff speak of population replacements, because of his public scorn of the notion of such events as embodied in the Out of Africa hypothesis. It is also puzzling to see the dual, apparently contradictory, fates of the Sangiran fossils in light of the new dates. Previously, when the fossils were thought to be about a million years old, they were cited by Wolpoff and his supporters as strong evidence for regional continuity in Australasia. The fact that the Sangiran fossils may be almost twice that age is "no problem," he now says. "It just extends the age of the Australasian clade." How is this to be reconciled with the suggestion that pre-Acheulean populations were wiped out by the technologically superior newcomers? "Well," he responds, "perhaps they weren't wiped out. Perhaps they mixed to some extent, sufficient to maintain certain anatomical features."[21]

If this were true, however, some ancient types of mitochondrial DNA would be present in modern populations, a molecular mark of ancient Java. In the more than four thousand people who have been tested to date, not a single example of ancient DNA has been found.

12

Headhunters at Ngandong

THE village of Ngandong, in central Java, is a mere speck on the map, barely a village at all, with nothing to attract visitors. Located amid the rolling terrain of the Kendeng Hills, Ngandong was once the head of a railway line that was used to transport ancient teak logs out of the surrounding forest and on to Madiun, where they were processed and then exported to Europe. But the track has long since been ripped up. The timber business still exists, but when logs are to be taken out these days, they go by truck along a narrow rutted road, which makes for a slow, tortuous trip. The few simple houses that constitute the village are widely scattered among the trees and poised high above the great Solo River, at a point where it makes a giant bow in its easterly course. Just six miles upstream is Trinil, where Eugène Dubois unearthed the first *Pithecanthropus* fossils at the end of the nineteenth century. In a country that teems with people and their agriculture, the terrain around Ngandong is an oasis of quiet, simply because its sandstone and limestone soils cannot support the cultivation of rice.

It was while he was savoring this rare solitude one evening after a long day of mapping the region's geology that, almost seven decades ago, C. ter Haar noticed something that was to make the inconspicuous hamlet of Ngandong a monumental name in the annals of anthropology. Ter Haar, whom his friend the Dutch paleontologist Ralph von Koenigswald described as "an amiable colleague, tall, thin and always good humoured despite his seven children,"[1] was employed at the time by the Dutch Geological Survey in Bandung, in western Java. His task was to produce a new geological map of the country. By the middle of 1931, his work west of the Solo River was almost complete, and he had based himself in Ngandong to continue his mapping eastward. On the evening of 27 August, after refreshing himself with a bath, ter Haar walked down to the river and sat on the bank to watch the sunset. Ever alert to the surrounding geology, he noticed a layer of sand and gravel in the steep bank about 60 feet above the river. Long ago the river must have flowed at that level, depositing the terrace that ter Haar saw, but over the millennia the water had slowly but inexorably sliced its way down to its present level.

Ter Haar scrambled to the ancient river terrace and began to poke around, as geologists love to do. He soon spotted a piece of fossilized bone protruding from the surface and tried to retrieve it with his hammer, but failed. He realized that the bone was probably quite large, so the following day he organized a team of his workers in a test excavation. The bone fragment that had first attracted his attention turned out to be part of an almost intact skull of a large buffalo. The excavation uncovered other bones, too, and stag's horns. Ter Haar quickly understood that he had stumbled upon what was obviously a rich cache of ancient bones, one that promised to be an important window onto the island's history. Two weeks later, the Geological Survey initiated a full-scale excavation, beginning at the precise spot where ter Haar had spotted the buffalo skull protruding from the terrace.

Ter Haar's find was in fact a rediscovery. A little more than twenty years earlier, G. Elbert, a geologist with the Selenka expedition, which was following in Dubois's footsteps upriver at Trinil, spent some time prospecting elsewhere and came across the bone bed at Ngandong. He judged the site to be geologically younger than the beds at Trinil, collected some material, and wrote a report. But because Elbert was considered to be "a man of excessive imagination,"[2] no one would believe him. Had she paid heed to her geologist, Margarete Leonore Selenka might have returned to Germany leaving a legacy more noteworthy than a scattering of beer bottles on the banks of the Solo at Trinil. As it was, the true value of the Ngandong site had to await the excavation by the Dutch Geological Survey, which extended for two years, beginning on September 12, 1931. The haul was astonishing: thousands of animal bones, twelve ancient human skulls or skull fragments, and two human leg bones. When anthropologists find one ancient human skull, it is cause for celebration. Finding two in one spot is regarded as extraordinary good fortune. Finding twelve was unprecedented at the time and in fact remains unmatched to this day. Unique. The Ngandong skulls came to be known in the vernacular as Solo Man.

As always, of course, the significance of the find was measured in terms other than sheer number. What kind of human species were the Solo people? And when did they live? These questions were immediately asked—and asked again, and again, and again. The Solo skulls eventually assumed an important place in ideas about how and where modern humans evolved, being contended by some to be the descendants of the early human inhabitants of Java, such as the Mojokerto child and the *Homo erectus* people of Sangiran, and also thought to be the ancestors of the people who first occupied nearby Australia, some 60,000 years ago. Our work on the age of the Solo fossils, and Susan Antón's study of the anatomy, which we describe in the following chapter, had a major impact on this important question.

Unearthing a Treasure Trove, and a Gruesome Speculation

Von Koenigswald had arrived in Java early in 1931, at the age of twenty-nine, some eight months before ter Haar discovered the ancient river terrace at Ngandong. Von Koenigswald's job was to study the animal fossils that were being uncovered at half a dozen locations on the island and stored in the Geological Survey in Bandung, which he described as "a beautiful city of villas."[3] Because of his interest in human prehistory, von Koenigswald became involved with the recovery and preliminary study of the Ngandong skulls soon after their initial discovery, and was to be associated with them in one way or another almost continuously for the following three decades and more.

The excavation, which started on September 12, was led by Mantri Samsi, *mantri* being the Indonesian word for a worker who is specially trained and can work without supervision. In order to excavate the bone-bearing terrace, Samsi and his men had to remove an eight-foot-thick layer of overburden, not a pleasant task in the hot, humid climate. The terrace itself, when they got down to it, was about six feet thick, with most of the bones concentrated near its base. The mantris very soon began finding fossils, which they dried in the shade and then wrapped in thin, tough Chinese paper that had been impregnated with glue to protect them for their impending journey, after having lain undisturbed for tens of millennia. Each fossil was numbered and the location and date of its unearthing recorded in a ledger. Just three days after the work began a relatively large, rounded skull was discovered, and was labeled "tiger." Two weeks later a second such rounded skull came to light, this time labeled "ape." Together with other fossils, the two skulls were packed into boxes, which were carried through the jungle to Ngawi, a distance of six miles, and then by train to Bandung, where the survey's director, W. F. F. Oppenoorth, was eagerly awaiting each new consignment of a petrified glimpse of the island's history.

Right from the start, the fossil haul had been impressive: buffalo skulls, the skull and lower jaw of a *banteng,* which is a species of wild ox still found on Java, well-preserved Stegodon jaws, and much more. When Oppenoorth saw the "tiger" and "ape" skulls he knew they had been mislabeled, and recognized them as human. Apparently, although the mantris were skilled in removing fossils carefully from the terrace deposits, their training in anatomy was less developed. Oppenoorth went right away to Ngandong to see the site for himself, where part of a third human skull had just been found. Obviously, Ngandong was turning out to be an extraordinary site.

Back in Bandung, von Koenigswald had the task of photographing the skulls for publication. When he first heard that human skulls had been found at Ngandong he fervently hoped that they would be *Pithecanthropus,* such as Dubois had found. In the 1930s most anthropologists thought that Asia—not Africa, as Darwin had written—was the birthplace of humankind. This view was based more on value-laden opinion than on fact or the kind of reasoning that Darwin had used. For instance, just three years before the discovery of the Ngandong site, Henry Fairfield Osborn had written: "The actual as well as ideal environment of [our] ancestors was not in warm forested lowlands . . . but in the relatively high, invigorating uplands of a country such as Asia was in the Miocene and Oligocene time—a country totally unfitted for any form of anthropoid ape, a country of meandering streams, sparse forests, intervening plains and meadow lands. Here alone are rapidly moving quadrupedal and bipedal types evolved; here alone is there a premium on rapid observation, on alert and skillful avoidance of enemies; here alone could the ancestors of man find materials and early acquire the art of fashioning flint and other tools."[4] Osborn's disdain for the assumed unsuitability of "the Dark Continent" for so noble an event as the evolution of humanity is obvious in the tone of his writing. The discovery of the Peking Man fossils in the late 1920s bolstered the Asia-centric view, as did the presence of *Pithecanthropus* in Java. Knowing that new specimens of *Pithecanthropus* would be very important for that anthropological position, von Koenigswald was very eager for this to be the type of human fossils that were coming out of Ngandong.

Von Koenigswald was apparently denied his fervent hope, however. "The first glance was enough to show me that—unfortunately—this was no new *Pithecanthropus,*" he wrote twenty-five years after the discovery. "The skulls were too lofty and too strongly vaulted, and the supra-orbital ridge—though massive and well developed—projected less sharply from the cranium." Nevertheless, he said, "it was a great discovery."[5] (It was only in the late 1940s that anthropologists came to agree with Darwin that humans originated in Africa, when the australopithecine fossils that had been found in South Africa were finally accepted as being human rather than ape, more than twenty years after the first one had been unearthed.)

Those first three skulls—the "great discovery"—soon proved to be only a beginning. A fourth skull was found on January 25, 1932, and a fifth, six weeks later. A total haul of five skulls was tremendously gratifying to Oppenoorth, but to that date he had not seen a single one prior to its removal from the ground.

This was a problem because, skilled excavators though the mantris were, an anthropologist's eye was necessary to determine whether the human skulls had become part of the bone bed contemporaneously with the other animal bones or had been incorporated later. If they were simply lying on or near the surface, for instance, they might have once been higher in the section—therefore of a younger age—and simply migrated downward as the deposit eroded. This would mean that the human skulls would be younger than the other bones with which they were now found.

When, on June 13, a sixth skull was spotted, excavation stopped, and a letter carrying the news was dispatched to Oppenoorth in Bandung. Oppenoorth immediately made arrangements to go to the site, accompanied by von Koenigswald, who was excited to make his first trip to central Java. After a journey by train, in horse-drawn carts, and finally on foot, Oppenoorth and von Koenigswald arrived at Ngandong five days after the skull had been found, its location having been marked by a few palm fronds. "We removed the fronds and ter Haar began to dig with his hands, while I took photographs," von Koenigswald later wrote. "Unfortunately, I was so excited that most of the shots were underexposed."[6] Ter Haar soon uncovered the skull, which was firmly embedded upside down in the bone-containing layer. Unlike the five previous specimens, this cranium had an intact underside; it proved to be "the most perfect specimen discovered at Ngandong," as von Koenigswald put it.[7] And the skull's situation in the bone bed—at least as important as the specimen's relative completeness—gave every indication that the skull was there at the time the bed was formed, rather than becoming part of it at a later date. This was crucial information for obtaining, by whatever means, a true age for the human skulls.

Work continued at the site until November of the following

year, when virtually the entire terrace had been removed. The final count was twelve human skulls or parts of skull recovered, and two bones of the lower leg. In all of them, however, the face was absent, and in many the underside of the cranium was also missing. All of them had once housed relatively large brains, an average of 1100 cc, smaller than the modern figure of 1300 cc but bigger than most *Pithecanthropus* brains. The first of the skulls to be found was badly damaged, having the appearance, von Koenigswald speculated, of having "been smashed, and by a mighty blow with a blunt instrument—perhaps a large wooden club."[8] The area around the fractures was black, probably having been stained by manganese during fossilization. To the mantris, however, the blackness was fossilized blood, forensic proof of prehistoric violence.

Von Koenigswald knew this to be untrue, but, he reports, he was unable to persuade the mantris otherwise. But von Koenigswald had a few flights of fancy of his own. Pointing out that in most of the skulls the base was missing, he drew a link with the practices of certain technologically primitive people of today. "If we examine the skull trophies of modern head-hunters, we find that here, too, the region of the foramen magnum is severely damaged," he wrote. "The head-hunter is not content merely to possess the skull, but opens it and takes out the brain, which he eats in order by this means to acquire the wisdom and the skill of the defeated foe. What we had found at Ngandong, therefore, were skull trophies."[9] In his book *Meeting Prehistoric Man,* which he wrote twenty-three years after the excavation of the Solo skulls, von Koenigswald described the Ngandong venture in a chapter entitled "The Head-Hunters of Ngandong," making it clear that he clung to the notion for a long time.

He thought of the bone accumulation at Ngandong as being the garbage of the headhunters who lived at this secure bend in the river; they placed the trophy skulls around the encampment, he wrote, to ward off evil spirits and threats from enemies.

Scenarios of prehistoric violence were once common in the anthropological literature, because most fossils are fractured in some way. But as the science of taphonomy—the study of the natural burial and fossilization process—has advanced in recent years, anthropologists have come to realize that most, if not all, of the damage seen in early human fossils is natural, not human-inflicted. In the case of Ngandong, the bone collection was not the remains of countless dinners of the Solo people but rather the natural accumulation of parts of animal carcasses washed along by the river and then deposited in the slower-moving waters at the bend. During such a process the carcasses would fall apart and the most fragile bones would break. The bones of the face and the underside of the human skull, very definitely fragile, were almost certainly smashed as they rolled along the riverbed before being deposited. The Solo people *might* have been headhunters, of course, but the fractured state of the twelve skulls from Ngandong is no proof of it.

Although the Solo skulls most cogently capture our imagination, the vast majority of the haul of 25,000 fossils from Ngandong was of other animals. Most of the bones were of various species of cattle and deer, but there were elephants, hippopotami, and rhinoceroses, many pigs, and a few tigers and panthers. The animals that lived at the time of the Solo people were the clear descendants of those that were around at the time of Trinil, that is, more than a million years ago. But there were fewer species in Ngandong times, an impoverishment in biodiversity that often happens on islands because of their isolation. As von Koenigswald noted when he first studied the fossils, their distinctly modern appearance suggested that they had lived relatively recently, probably in the late Pleistocene.

He also pointed out that at a site contemporaneous with Ngandong the remains of a heron had been found, of a species that still lives in China but that does not venture farther south than the Yangtse River. "Only a deterioration in the climate could have driven this bird so far south," he said. "This makes it probable that the Ngandong levels coincide with the height of the last Ice Age, so that Solo man is contemporaneous with our classic European Neanderthaler."[10]

From the evidence of the Ngandong fossil animals and the wayward heron, von Koenigswald drew the inference that the Solo people probably lived somewhere between 30,000 and 150,000 years ago, in terms of hard numbers. There was no way to pin down the date more closely.

A Long Odyssey

Von Koenigswald remained in Java until 1946, spending most of his time trying to unravel the history of the island as reflected in the kinds of animals that had lived there during the past 2 million years. Along with looking at fossils discovered by others, he made important discoveries of his own in the form of the *Pithecanthropus* finds from Sangiran that we mentioned in chapter 5, and he collaborated with the German paleontologist Franz Weidenreich, who was working in China between 1937 and 1941, studying the remains of Peking Man, or *Sinanthropus.* The two men, remember, eventually came to the conclusion that *Pithecanthropus erectus* and *Sinanthropus pekinensis* were the same kind of human, which later came to be called *Homo erectus.* Early in their collaboration Weidenreich and von Koenigswald had quickly formed a close friendship and professional relationship, and the unfolding circumstances of the age conspired to bring the two men together in a way that neither could have anticipated, with the Solo skulls playing an important part as well.

With World War II already roiling in Europe in mid-1941, Weidenreich judged it wise to leave China for America, which he did just months before the Japanese occupation of Peking (now of course called Beijing) later that year. He took casts of all the Peking Man skulls with him to the American Museum of Natural History, in New York, a providential move because the original fossils disappeared from Peking under mysterious circumstances at the beginning of the war there, never to be seen again. Meanwhile, in Java, von Koenigswald was hurriedly trying to safeguard the Javan fossils by making casts and substituting them for some of the originals. "The casts were extremely well made and to lay eyes almost indistinguishable from the originals," he later said. "We had mixed finely ground brick dust with the plaster of Paris, so that even in the event of injury the break would remain nicely dark, as in the genuine fossil. We switched the skulls, so that if the contents of the safe should one day vanish eastwards a few original pieces, at least, would remain in the country."[11]

Several people became involved in an often comical conspiracy to prevent the precious relics from falling into enemy hands.[12] As a result, all the Javan fossil humans survived the war unscathed, despite the fact that one of the Solo skulls had been sent to Japan as a birthday present for Emperor Hirohito. (It was safely retrieved after the war.) Von Koenigswald experienced a more difficult time than his fossils, suffering terribly as a prisoner of war under the Japanese between 1942 and 1945. When he was released he wrote to his friend Weidenreich, telling him of his wartime travails and giving him the news that the Javan fossils were safe. Weidenreich, who had believed his friend dead, was much relieved to hear from him, and immediately undertook to have von Koenigswald join him in New York. Von Koenigswald arrived in September 1946, accompanied by his wife, his six-year-old daughter, and the most important Javan human fossils, including the Solo skulls. Of the 25,000 animal fossils from Ngandong, however, there was no trace, save for a few that were in von Koenigswald's personal collection, the rest having disappeared under circumstances that no one has ever explained. They may still exist in some paleontological warehouse in Java, but if they do, they do so anonymously.

Von Koenigswald had intended his New York sojourn to be just a year, but ended up extending it by six months. "Having been a prisoner of war under the Japanese, my health was so poor that it soon became evident that one year was too short a period in which to recuperate," he later wrote. "In addition, the material was so rich that it was equally obvious that it would be impossible to complete our studies within that period."[13] For those eighteen months von Koenigswald and Weidenreich worked intensively on the Javan fossils, with von Koenigswald concentrating on the study of jaws and teeth, his professional love, and Weidenreich on the Solo skulls, which had received relatively little close investigation before that time; there had been no publicly available detailed descriptions of the fossils, of the sort that other anthropologists need if they are to analyze them at a distance. It was, by all accounts, a joyous and productive time for both men.

In the spring of 1948, von Koenigswald, his immediate work complete, left New York to take up a post at the University of Utrecht, Holland, of course taking the fossils with him. Weidenreich remained in New York. Having made beautiful casts of all the specimens he needed, he was deeply immersed in preparing a complete scientific description of the Solo skulls, for which anthropologists had been waiting eagerly for twenty years. On July 11 of that year, Weidenreich died unexpectedly at the age of seventy-five, the monograph incomplete. Poignantly, the manuscript stops in midsentence, during the description of the anatomy of the underside of the Solo Man cranium. Harry Shapiro, head of anthropology at the American Museum of Natural History at the time, described Weidenreich's incomplete "magnum opus" as having begun "on a Michelangelesque scale, rich in significant detail and masterly in conception."[14] The manuscript was published by the museum in 1951, as had originally been agreed with von Koenigswald, and with a foreword by him. Incomplete though it is, and finishing in midsentence as the manuscript did, the monograph remains one of the most important documents in anthropological literature, and without doubt a monument to and example of fine anatomical description.

Von Koenigswald spent a productive twenty years in Utrecht, publishing 145 scientific papers, a towering figure in the maturing science of paleoanthropology. When he "retired" in 1968, he established himself at the Senckenberg Research Institute and Natural History Museum in Frankfurt, where he had been promised funds and facilities for setting up a center for curating and studying the Javan fossils. A decade later, just four years before he died, and with his health beginning to fail, von Koenigswald returned the Solo skulls and the Mojokerto child's skull to Java, to the safekeeping of Teuku Jacob, with whom he had enjoyed a warm professional relationship for more than a decade.

The odyssey of the Solo people—from life to death, from the Old World to the New World, and back again—was finally over.

Who Were the Solo People?

A cache of a dozen skulls, such as those from Ngandong, is invaluable to anthropologists, not least because it allows for some insight into the anatomical variation that existed within an ancient population. As we will see in the next chapter, measuring anatomical differences among fossil individuals and judging their biological significance has become a big topic in paleoanthropology, one that speaks to our sense of who we are in the world. Beyond this, however, the Solo people have come to represent a major pillar supporting one of the two principal models for the origin of modern humans, that of the multiregional evolution hypothesis discussed in chapter 11. *Who* the Solo people were—that is, what species they belonged to—and, as important, *when* they lived turns out to be a major test of that hypothesis.

Ever since their discovery, the Ngandong fossils have been the subject of a cottage industry asking the "Who were the Solo people?" question, and a glance at the anthropologists' bible, *Guide to Fossil Man,*[15] shows what a range of opinion there has been. Starting with Oppenoorth's original published opinion in 1932 and working toward the present, and indulging for just a minute in a little nomenclatural immersion, we find that the Solo skulls have been dubbed: *Homo (Javanthropus) soloensis, Homo soloensis, Homo primigenius asiaticus, Homo neanderthalensis soloensis, Homo sapiens soloensis, Homo erectus erectus, Homo erectus,* and most recently *Homo sapiens*. Clearly, nothing is absolute in anthropology, especially so in this case. This unusually large blizzard of formal names reflects as much the shifts in thinking about recent human evolution as any new understanding of the anatomy of the skulls. In anthropology, which hypothesis an individual favors tends strongly to influence his or her interpretation of data, perhaps more so than in any other science.

The issue with the Solo people comes down to this: Were they primitive, in the sense that *Homo erectus* was primitive, that is, among other things, possessing a smaller brain than modern humans, thick cranial bone, and a long, low cranium that sported prominent brow ridges, and were they behaviorally less competent than modern humans? Or were they much closer to *Homo sapiens* than to *Homo erectus,* anatomically and behaviorally? Early opinion was very much toward the latter, despite the Solo people's thick cranial bone, low cranium, and prominent brow ridges. In his 1932 paper Oppenoorth said that the skulls were a lot like those of Neanderthals, although they were different from them, which is why he chose the name *Homo (Javanthropus) soloensis.* He later recanted on the suggested resemblance to Neanderthals, apparently after being browbeaten by Eugène Dubois, who insisted that the Solo people were virtually completely modern.

Von Koenigswald and, later, Weidenreich also thought there were echoes of Neanderthal anatomy in the Solo people, though Weidenreich somewhat less so than von Koenigswald. Despite the formal name that Oppenoorth had proposed, said von Koenigswald, "Solo man proved to be a Javanese Neanderthaler."[16] Although Weidenreich considered the Solo people to be somewhat more primitive than Neanderthals, both he and von Koenigswald were essentially saying that the people of Ngandong were distinctly more advanced than *Homo erectus.* One of the most seductive sirens calling toward this conclusion was the relatively large size of the brain boasted by the Solo people. As observers of ourselves and our ancestors, we humans are especially obsessed—some would say excessively obsessed—by big brains.

Weidenreich went on to develop the notion that there was a continuous evolutionary chain stretching a million years or more in Southeast Asia. "There is now an almost continuous phylogenetic line leading from the *Pithecanthropus* group through Homo soloensis to . . . the Australian aboriginal of today," he wrote in 1945.[17] In other words, the Solo people were the descendants of *Pithecanthropus,* or *Homo erectus,* and the ancestors of the earliest Australians. From early on, then, the essence of the multiregional evolution model—that of regional continuity—was expressed by Weidenreich, with Southeast Asia the prime example. In this scenario, the Solo people were considered to be anatomically intermediate between *Homo erectus* and *Homo sapiens.* A relatively recent age for the Ngandong fossils was therefore acceptable, something close to the high end of the theoretical range—that is, more than 100,000 years old.

Then, in 1980, A. P. Santa Luca, an anthropologist at Yale University, published the first major reassessment of the Ngandong skulls since Weidenreich's 1951 magnum opus. Santa Luca came to believe that the Solo people were not as "advanced" as had been generally thought—that is, that they were not close to the *Homo sapiens* state—and said that they should be "accepted as a *Homo erectus* group."[18] But this introduced a disconcerting factor: if the Solo people were *Homo erectus,* then they couldn't be as young as most people had assumed, based on the known age of the other animals in the river terrace at Ngandong, because everyone "knew" that *Homo erectus* had disappeared at least a quarter of a million years ago. He therefore suggested an age of closer to 400,000 years ago for the Solo skulls. Santa Luca's work sowed the seeds of a curious dance that was, and continues to be, played out over the identity and age of the Solo people. Those people—principally geologists—who see signs of a young age in the river terrace deposits assume that the skulls must be fairly advanced, certainly not *Homo erectus.* Others—some anthropologists—who see signs of primitiveness in the skulls assume that the river terrace deposits must be old, or at least that the human skulls themselves must be old even if the animal bones alongside them are not. Different truths are in the eyes of different beholders.

Meanwhile, the multiregional hypothesis of the origin of modern humans began to be shaped into its modern form, the intellectual descendant of earlier incarnations such as

Weidenreich's. To reiterate, the proponents of the hypothesis, chief among them being Milford Wolpoff of the University of Michigan and Alan Thorne of Australian National University, argued the following: that each population in every part of the Old World today is descended from populations of *Homo erectus* that were established in those same geographical locations long ago; that these deep genetic roots explain the anatomical characteristics that distinguish modern populations from one another, characteristics that go back to their *Homo erectus* ancestors; and that there was a small degree of gene flow—that is, interbreeding—among the different populations.[19] In other words, populations were sufficiently isolated from one another to maintain their local anatomical features, or regional continuity; but they were sufficiently in contact with one another to maintain all human populations as a coherent species. That was, and is, the argument.

And whenever they have presented their argument, whether in professional journals or in more popular outlets, Wolpoff and Thorne have consistently adduced the evidence of Southeast Asia to bolster their case. Joined by Wu Xin Zhi, of the Institute of Vertebrate Paleontology and Paleoanthropology in Beijing, Wolpoff and Thorne opened their 1994 hypothesis-establishing paper as follows: "The east Asian hominid fossil sequence presents an unequaled opportunity for the development and testing of hypotheses about human evolution."[20] In a popular British weekly science magazine they explained their position this way: "In each region of the world, we have uncovered links that tie living populations to their local antecedents, whose remains are preserved in the fossil record. The most striking evidence comes from Asia." They went on to describe such a putative series of links in Indonesia and Australia. "The sequence starts around a million years ago, with the remains of Java Man—a representative of the hominid species *Homo erectus* discovered in 1891—and ends with the remains of Australians dated at around 10,000 years ago."

This supposed "sequence" is in effect just three data points: the earliest, thought at the time to be close to a million years ago, that is, the Sangiran and Mojokerto *Homo erectus*; next, the Solo people, which the article's text said were "around 100,000 years" old, although Wolpoff has said in the latest (1999) edition of his textbook that he prefers an age closer to 250,000 years; and the early fossil Australians. Most scientists would be uncomfortable with a "sequence" that is a vast stretch of empty time punctuated by three tiny islands of data, and even more uncomfortable with the idea that this is the "most convincing evidence" of a hypothesis.

As we explained in chapter 11, our new ages for the Sangiran and Mojokerto fossils made such a sequence even less tenable because they effectively doubled the time over which the sequence was supposed to endure. As we also mentioned there, Wolpoff tried to save the multiregional hypothesis by suggesting that perhaps the Sangiran and Mojokerto people were not ancestors of Australians after all, but merely representatives of an early, failed migration into Southeast Asia. This new position left the rest of the "sequence" intact, however, with the Solo people being the ancestors of the earliest Australians, anatomically modern *Homo sapiens* people who, according to archeological remains, arrived by boat on the island continent some 60,000 years ago.

13

Facing the Inescapable

DURING Garniss's and my first trip together to Java in the fall of 1992, Jacob had expressed the wish to "finish Ngandong" in his declining days. He had organized excavations there between 1976 and 1980, the first since Oppenoorth's time, and had uncovered many animal bones as well as some human bones, including two partial crania. But then the money ran out, and he had to stop. The age of the human fossils was still up in the air a decade later, even though Dutch scientists in the 1980s had dated some of the animal bones at around 100,000 years old. Jacob asked us whether we would be willing to apply our geochronology skills to Ngandong so that the age question could be settled once and for all. We agreed, thinking it would be straightforward. We'd collect some samples for dating, do some ages, write a paper. That would be it, we assumed. Boy, were we wrong.

During our brief visit to Ngandong, after our excursion to the Mojokerto site in early September 1992, we looked for pumice at a place close to the site of the 1930s excavation, hoping for suitable volcanic minerals to date. There wasn't much at all, just pea-sized pieces. "It took me hours just to collect the stuff," remembers Garniss. "Just a few grams in a little plastic bag." We knew that dating the mineral based on the potassium/argon approach was going to be problematic, even with the very sensitive single-crystal laser-fusion method of argon-40/argon-39 dating, for two reasons. First, as elsewhere in Java, there was very little potassium in the volcanic minerals at Ngandong, which meant that the amount of radiogenic argon produced would be low, making it tricky to measure it. Second, if the rocks were as young as people believed—that is, between 400,000 and 100,000 years—then there would have been very little time elapsed for the generation of radiogenic argon, thus compounding the measurement problem. But we considered it worth the small amount of effort it would take to collect pumice and run it through a quick dating process.

It took about a year before I finally got around to working with the poor samples of pumice that we had brought from Ngandong. The first sample came out at around 450,000 years, so it looked as if the Solo people might be old after all. The second, however, was much younger, close to 100,000 years, but with a lot of uncertainty in that figure. This disparity of ages, along with the rarity of pumice in the alluvial deposit, indicated that the pumice wasn't primary, wasn't part of the original deposit, but had become incorporated from other ash layers into the Ngandong river terrace. I was dating pumice all right, but who knew where it came from originally? Not sure what to do next, we put the work aside.

Five months later I ran into Clark Howell at the February 1994 meeting of the American Association for the Advancement of Science, in San Francisco, the one at which we had announced our new age for the Mojokerto child and the Sangiran fossils. It was lunchtime, and Clark was sitting at a table in the cafeteria, so I joined him and we exchanged greetings. Then Clark said to me, indicating the short, gray-haired man sitting next to him, "Have you met Henry?" I hadn't, but was delighted to learn that he was Henry Schwartz, the head of one of the world's preeminent labs for dating relatively young specimens, at McMaster University in Hamilton, Ontario, using newly developed techniques such as electron spin resonance and uranium series dating. I told Henry about the situation at Ngandong, saying that I thought it was an interesting challenge. I explained that the geological setting looks young. If the terrace was as old as many anthropologists contended, it seemed to me that with the amount of water going down the river, particularly in the rainy season, the river course would have been cut much deeper and wider. Would Henry be interested in a collaboration? We were planning to return to Java in the fall of that year, I told him, and we could collect some relevant material—teeth and surrounding soil—on which Henry could do electron spin resonance. Henry knew the Ngandong story, of course, and was intrigued with the possibility that the bone bed was much younger than was typically assumed. Yes, he said, he would be happy to join in the venture.

Like potassium/argon dating, electron spin resonance dating, or ESR for short, depends on detecting the products of radioactive decay. In this case, the principal radioactive element is uranium-238, which, when it decays to an isotope of thorium, releases a pulse of energy in the form of gamma rays. The crystalline structure of tooth enamel makes teeth a suitable target for ESR dating, in the following way. When an animal dies in the natural environment and its bones become buried, its teeth contain no uranium, so they have not been exposed to this radioactive decay process. This particular radiogenic clock is therefore naturally set at zero when the animal starts to become part of the fossil record.

As time passes, gamma rays from the decay of uranium-238

in the soil surrounding the teeth, and from uranium that starts to insinuate itself into the crystal lattice structure itself, bombard the atoms that make up the crystalline enamel. From time to time a gamma ray burst knocks an electron out of its orbit around an atom and propels it to a higher energy state. These errant, high-energy electrons can drift around the lattice but usually fall back to what physicists call the ground state. However, a small proportion of them get trapped in the lattice at an intermediate energy level. The more time has passed, the more intermediate-energy-level electrons there will be in the tooth.

The ESR clock is based on a method of detecting the quantity of intermediate-energy-level electrons there are in a target sample. The electrons act like minute magnets, and when the tooth is exposed to a strong magnetic field they become oriented like compass needles. This orientation can be flipped—that is, north to south, and south to north, so to speak—by reversing the magnetic field. The more electron "magnets" there are in the sample, the higher the level of alternating magnetic fields required to flip them all. This is a signal of how old the tooth is, because a greater quantity of intermediate-energy-level electrons indicates a longer time since the animal died. Obviously, the amount of uranium there is in the soil will influence the rate at which electrons are bounced out of their ground state and into a higher state. ESR dating therefore requires the collection not only of a tooth—or teeth—but also of the soil in which it was entombed, so that the uranium concentration can be measured and worked into the equation for calculating age.

We planned to return to Java in August 1994, when, among other things, we would collect teeth and soil samples from Ngandong. Meanwhile, Garniss wrote to Jacob and asked him to send one of the teeth that had been collected during their excavations. In the absence of associated soil, this would at least allow Henry and his colleague Jack Rink to do a "dry run," or reconnaissance date, to find out if there was sufficient uranium in the fossils to make ESR dating possible. There was. "The teeth that you sent us appear to be excellent for ESR and uranium series dating," we wrote to Jacob on August 5, 1994. "We have just heard that [the preliminary results] indicate a very young ESR age." By very young we meant possibly as little as 100,000 years, which, we wrote, "is much younger than you or we believe, but given this new data, we need to be sure that we cover all possible sources of error. This could be the most exciting controversy yet!" Despite there being suitable teeth in collections that had been made earlier at Ngandong, we insisted on doing the final dating on specimens that we ourselves had uncovered from the bone bed, so that there would be no doubt about where they had come from.

We arrived at Ngandong in the late morning of August 26, 1994, accompanied by Agus, but we were not yet ready to get to work because we had brought with us some supplies that Sharie Shute, Garniss's assistant, wanted us to deliver to the school, including a blow-up globe, maps, and pencils. On an earlier visit she had established a relationship with the Ngandong school principal and her charges, and Sharie wanted to augment their meager stock of learning materials. The Ngandong schoolhouse was right where it had been on the previous visit, but this time there were no schoolchildren in sight. It turned out that the school had been closed down because there were too few children in Ngandong, and now the children had to go to a neighboring village. It was only a few miles away through the teak forest, but we discovered that the drive was worse than the one into Ngandong, which had been bad enough.

When we got there, after what seemed like hours, we were told to wait for the school principal in the mayor's house.

Within minutes every window was full of faces just staring at us from the outside. It was obvious that not many Westerners had been to this village. In one corner of the room we were waiting in was something that looked like an operating table, with stirrups at one end. We had no idea what it was until Agus explained that it was part of the Just Two policy: any woman who had more than two children was taken off to the mayor's house to receive a free IUD, like it or not.

By the time we got back to Ngandong it was already four o'clock in the afternoon, later than had been planned and much later than was ideal, given how much we had to do. Agus had earlier sent word ahead, asking that the overburden be removed at one location at the site, so that all I would have to do would be to dig small test pits in search of teeth. But that hadn't been done, so the work facing us was even more than we had expected. I eventually got down to the bone bed and exposed about a square meter, into which I cut three pits, each just a few inches across. Remarkably, in each of them I found a tooth, two antelope and one elephant—which demonstrates either that this is a very rich bone bed or that I have the paleontological equivalent of a green thumb, or both.

Garniss, meanwhile, was squatting on the dirt floor in a small house nearby, preparing dosimeters over a propane gas stove in almost total darkness. Dosimeters are often necessary in ESR dating because teeth in the ground are exposed to gamma radiation not only from uranium in the soil but also from cosmic rays that are striking the Earth from outer space. A dosimeter is a simple piece of equipment in the extreme, just an inch-long copper tube packed with radiation-sensitive material. Placed in the ground right where the sample fossil has come from, the dosimeter measures the total radiation at that spot, both terrestial and extraterrestrial, over a period of about a year. Geochronologists can then take into account the cosmic ray radiation's contribution to the ESR signal. But in order to set them to zero, dosimeters have to be heated and be in the dark immediately before they are buried.

When I had finished my mini-excavations and Garniss was done with his work, we had to figure out a way of burying the five dosimeters so that the young children who had been watching the whole process with great fascination didn't just dig them right up again as soon as this strange group of foreigners left. In the end we got some bamboo canes, tied the dosimeters to the ends, and inserted them into the bone bed at the right level but so that their position would be hard to detect. Then, to keep overinterested fingers from prying, we left a little money for a local family to watch over the site. Everyone was relieved to be out of Ngandong by five-thirty, ready for a long drive to Madiun.

The bamboo ploy seemed to work, because a year later all but one of the dosimeters were safely retrieved. As it turned out, they weren't necessary for the final calculations. The level of uranium in the Ngandong bone bed soil was so high that, by comparison, the extra radiation from outer space was insignificant.

We also collected teeth and soil samples from the sites of Sambungmacan, about thirty miles upriver from Ngandong and at the same geological horizon, that is, the same age as the Ngandong fossil bed, and Jigar, a fossil vertebrate site located just downstream from Ngandong. Jacob had excavated here, too, in the late 1970s, and had found many fossils, including one human cranium. Obtaining an ESR age of this site would provide a test of whatever age we obtained at Ngandong. If they proved to be the same, we could be more confident of our Ngandong results. If the dates were different, then we could start to worry.

Facing the Impossible

During 1995, in their lab in Hamilton, Ontario, Henry and Jack started to get dates on the samples that we had collected on the trip, and we were happy to hear that the Ngandong, Sambungmacan, and Jigar dates matched. Meanwhile, however, Susan Antón was deep into her study of the anatomy of the Solo people. Their separate conclusions—Henry and Jack's on the age of the Solo people, Susan's on who they were—seemed to be moving inexorably in incompatible directions. Something would have to give, that was clear.

With the soil samples and the new tooth specimens, Henry and Jack were making what were surely more accurate age estimates than had ever previously been possible, even before the dosimeters were collected. If 100,000 years old had been considered too young to be believable, then 50,000 years old seemed out of the question. But 50,000 years old, or something close to it, was what was emerging from the calculations.

The ESR approach is based on highly complex chemical and physical processes in nature, such as the rate of diffusion of uranium into the crystal lattice, and there are many technical traps for the unwary. In science, extraordinary claims have to be supported by extraordinarily solid evidence. The claim we were going to make was without doubt extraordinary, because, as I wrote to Henry in an e-mail on October 8, 1995, "many will argue that 50,000 years is too young for *Homo erectus.*" So, in addition to scrutinizing every possible interpretation of the ESR data, Henry decided that a second line of investigation should be followed, that of uranium series dating.

Like ESR, uranium series dating depends on the radioactive decay of, among other things, uranium-238 to thorium-234. But unlike ESR, U-series dating, as it is usually known, measures the decay products themselves, not their effects on something else. By measuring the ratio of the various isotopes of uranium and thorium, it is possible to judge how long the decay process has been going on in the sample, and thus how long ago the animal died. U-series dating can be done on fossil bone, not just on teeth as with ESR, but in this case teeth were what were available. Henry's U-series results on the Ngandong and Sambungmacan samples were stubbornly the same as the ESR results, ranging between about 30,000 and 50,000 years.

On November 10 I e-mailed Susan, who is based at the University of Florida in Gainesville: "Well, to make your head even more fuzzy . . . get this . . . I just got the U-series age back on one of the Ngandong teeth (I am doing three). The age is 38,000 years. . . . How the hell are we going to explain *Homo erectus* in Java between 38 and 32 kya [thousand years ago]? Are you sure you don't want to make the Ngandong hominids *Homo sapiens*?" The more Susan had studied the Ngandong skulls, the more convinced she became of their proper identity. "What you see is a pumped-up *erectus* skull," she explains. "It's the same shape—bigger, yes, but the same shape. You don't see the globularization you get in *Homo sapiens.* If you look at the shape of the temporomandibular [jaw] joint, the shape of the cranial base, the place where the cranium is widest—what you see is all characteristic of *Homo erectus.*" The anatomical differences that do exist between Ngandong and earlier *Homo erectus* are simply a consequence of Ngandong's larger brain, she says.

"When I was asked if I didn't want to make Ngandong *Homo sapiens,* I'd say, 'I'll go back again and check, if you like!' " she recalls. "I did that once. But no, I was sure it was *erectus.* And for the first six months after the young dates started to come out I thought there must be something wrong with them."

Jacob thought there must be something wrong with them, too. When I first told Jacob that the age was beginning to look very young, he said that the skulls might be as old as 400,000 years, which he obviously preferred, and added that a Japanese colleague using gamma ray dating had obtained a not yet published result of at least 350,000 years. But he was a bit mysterious about it all, not revealing who the scientist was. When I told Henry about it, he did some sleuthing and found out that the scientist was C. Yokoyama, a Japanese geochronologist working at the Museum of Man in Paris. Yokoyama had collaborated with Christophe Falguères, a French geologist, and obtained the older age two years earlier using a technique known as gamma ray spectroscopic analysis. Like ESR and U-series dating, this too depends on the decay of uranium-238, but unlike the other two techniques, it seeks a direct measure of the gamma ray radiation that occurs during the decay processes. It has the great advantage that you can date the object of interest—in this case a skull—and not something associated with it. In the realm of anthropology, however, it was virtually untested at the time. And although it should work in principle, it may be difficult to produce an accurate answer with an object as large and irregularly shaped as a human cranium.

The technique requires only the most basic of equipment. The target sample—in this case, it was the first of the Ngandong skulls found in September 1931—is placed in a box lined with lead bricks, and a detector picks up the profile of radiation that emanates from the sample's surface. The radiation profile is generated by the ratio of the various isotopes in the fossil, and, remember, this ratio gives an indication of the age of the fossil. So, while U-series dating gets an age by measuring the isotope ratio chemically, gamma ray dating does it indirectly, by detecting the radiation coming from the decay of those isotopes. In any case, when in May I learned from Henry that the gamma ray work had been done in Paris, I contacted Susan, who just happened to be there, and asked her to talk to

Yokoyama and Falguères about the work. She did, and found that there had been sufficient concerns about the reliability of the date—the skull had been exposed in the lead-lined box for too short a time, for instance—that they had not felt confident about publishing the result at that point.

Despite the uncertainty surrounding the gamma ray procedure, when we first heard about it we wanted to address it somehow. There was the worry that if we went ahead and published our results with Henry and Jack, we might face criticism because this other work contradicted our conclusions. But the French work was unpublished, and so we did not feel we could refer to it—to try to refute it in some way—in our own article. (By mid-1999, the only mention of the work in the scientific literature has been a short abstract from a conference that says the skull is "older than 200,000 years.") A better route would be to address the issue head-on and conduct more tests of our own. If we could obtain some bone, we could do U-series dating directly on the human material. In September 1995, we asked Jacob if we could drill some material from the human leg bones at Ngandong so that we would get an age from the fossils themselves, not just from animal teeth associated with them. Jacob said yes. Each time we tried to put the plan into action, however, Jacob made some excuse, and nothing happened.

As 1996 wore on, the young ESR and U-series dates tenaciously remained young, no matter what complicating issues Henry and Jack took account of, no matter how many different kinds of analysis they did, no matter how many tests they did. And the implication loomed ever larger: *Homo erectus* had survived in Java until perhaps 53,000 years ago, and maybe as recently as 27,000 years ago—in either case, far more recently than anyone had believed possible.

We all knew we were going to be attacked for those results. For Susan it was, "What career? As an anthropologist, how am

I going to say *Homo erectus* lived 50,000 years ago and expect to survive professionally?" Jack became distinctly nervous at going public with so maverick a proposal. But Garniss and Henry loved it. "If we were sure about the dates," says Garniss, "we knew we would just have to publish and be damned."

Early Reactions

With no progress in sight for obtaining some human bone from the Ngandong specimens on which U-series dating could be done directly, we began to write the paper on our work dating the animal teeth. We just couldn't wait any longer. Ever more honed versions of the manuscript were passed between Berkeley, Hamilton, and Gainesville, in a challenging dance that balanced the need to supply sufficient information so that readers could appreciate all the safeguards that had been taken with the dating procedures, and the need to abide by the severe length limitations that the journal *Science* puts on papers. "It is essential that we indicate the degree to which we have cross checked these results and avoided any possible pitfalls in the ESR dating," Henry said in an e-mail to me on May 14, 1996. "This will take at least one page of text, carefully whittled down from the usual massive outpouring of technical details that we love to spout."

The task was completed by early June and mailed to *Science*'s editorial office in Washington, D.C. Then I left for fieldwork in Mongolia and continued on to Java, Garniss departed for Turkey, and Susan headed to various stopoffs in Europe on her way to Java. Everyone was relieved to be able to concentrate on other matters for a while, but we knew that the inevitable brouhaha was not far away.

All papers submitted to *Science* are subject to peer review, which involves the paper being sent to other experts in the field for their opinion and then revised by the authors. This is an awkward limbo time—often lasting months—for a scientific paper, because, until it is actually published, officially it doesn't exist. People are not supposed to talk about it and certainly not supposed to write anything about it. In reality, however, reactions often begin to travel along the academic grapevine, particularly if the paper is in any way controversial, as this one certainly was. "Well the attack on the young dates has begun," I e-mailed Henry on November 11, just two weeks after the paper had been accepted by *Science.* "Rainer Grün was visiting Chris Stringer the other day and told him that Falguères & co. in Paris has gamma ray dated a Ngandong hominid and it was . . . at least 300,000 years old, so Rainer still thinks the hominids could be reworked. I predict that these dates will appear as a rebuttal to our work as soon as it is published." Grün, a German geochronologist working in Australia, had seen an early version of the paper and knew that Stringer, at the Natural History Museum in London and the chief proponent of the Out of Africa hypothesis, would be interested in knowing about it and about the possible weaknesses of our work.

Two weeks later, Henry received an e-mail from Alan Thorne, one of the two chief proponents of the multiregional evolution hypothesis and therefore no enthusiast for young Ngandong dates. "It will not surprise you to learn that I have received a variety of copies of the forthcoming *Science* paper on the dating of the Solo deposit," Thorne began. After saying that he found the science "very professional, as far as it goes," he went on to criticize the paper on two grounds. The first was that older dates were available and that by failing to mention them we were guilty of "hiding information." Second, he wrote, the human fossils at Ngandong were very different in appearance from the others—the human bones were darker in color and hard, the animal bones lighter and more fragile. This would imply that the two groups of fossils had different histories, and that perhaps the human fossils were much older. (It is certainly true that geological processes can erode a bone bed, so that the fossils it once contained can get redeposited in younger layers as they are being formed, making the migrated fossils appear to be younger than they are. If Falguères's date of 400,000 years for the first Ngandong skull was correct, then something like that must have happened.) Thorne also said that the paper was an obvious attack on the multiregional evolution hypothesis, and that it looked as though Garniss and company were trying to "curry favor with the Replacement [that is, Out of Africa] people."

I was irritated, particularly with the suggestion that my colleagues and I had been "hiding information" by not mentioning Falguères's work. "Yokoyama has told Henry that he could not use his dates in our publication," I responded to Thorne. "Given that I do not have permission to publish and discuss [the Paris group's] data, I don't see how I could have handled it differently. It would have been unethical for us to publish their ages and discuss why we think they are wrong before they even have a chance to publish not only their dates but their methods as well." We were not "hiding information," as Thorne suggested, but simply following the unwritten rule of science that you cannot cite unpublished work without permission.

I also strongly disagreed that the human and other animal fossils looked different in ways that implied different histories. "The nonhominid Ngandong fauna . . . which I looked at is not friable or different in appearance from the hominids," I wrote to Thorne. The human fossils did look different, I said, but this was the result of "shellac and polishing and casting over the years in the collection." As for the suggestion that the paper was meant to "curry favor with the Replacement people," I ignored it. After all, my colleagues and I had no vested interest in the outcome of the debate. We simply wanted to get good dates that might help settle it.

Paleontologists who are interested in the age of their fossils always have to worry whether the fossils were actually buried in the layer from which they are later excavated. A whole series of geological processes can conspire to rework fossils so that they finish up in new sedimentary layers, looking sometimes older, sometimes younger. It was ironic that we stood accused by some of claiming "too old" a date for the Mojokerto child and then "too young" a date for Ngandong. And in fact it was extremely unlikely that the Ngandong skulls were reworked. Having one skull reworked is possible. Having two, less likely. As for the odds of having a dozen skulls reworked and finishing up in the same spot—very, very small. And Sambungmacan bolstered our confidence: we had gotten the same age there, too. If, as people were arguing, the Solo skulls were really 350,000 years old, then you'd have to say that these old skulls were reworked into not just one site but two, which is even harder to explain than how a dozen old skulls got reworked into 40,000-year-old beds at Ngandong alone.

Fossils that get reworked usually suffer on the journey, later showing signs of scratching and abrasion and often having become even more fractured than is usual in the fossilization process. The Ngandong fossils do not fit that pattern. "You just don't see the kind of damage you'd expect to if they had been reworked," says Susan. "And Solo six has an almost intact cranial base. That wouldn't happen with reworking."

One person who was present when the Ngandong fossils were being recovered, and who had an opportunity to inspect them closely before time and differential treatment might conspire to make the human bones look different from the rest, was von Koenigswald. He had this to say about the matter in his introduction to Weidenreich's 1951 monograph: "That the human remains from Ngandong were contemporary with the laying down of the sediments is evident from their preservation, which is exactly the same as in the numerous animal bones found in the same deposit."[1]

Critics of the same-age position point out an anomaly with the Ngandong collection: Why, they ask, should there be all kinds of bones of other animals at the site—legs, arms, and vertebrae as well as skulls, for instance—but of humans the remains are overwhelmingly skulls, with just a couple of leg bones? Something must have been sorting them somehow. That's true, but that something might well have been the excavators who were unearthing the skulls rather than the processes that were involved in burying them in the first place. "The kinds of hominid material that were retrieved, the skulls, were the kinds that were easily identified as such, big enough to be brought out to be shown to someone who could identify it," Susan suggests. "When Jacob excavated in the 1970s, he came up with a lot of postcranial material, parts of lower limbs and so on, that hadn't been seen in earlier excavation. It is therefore very probable that there was a lot of fragmentary postcranial human material mixed in with the faunal material, but not identified as such." That huge original haul of 25,000 fossils is missing, so we will never know for sure whether that is the explanation for the apparent sorting of human fossils.

The Science Paper Is Published

Bearing the title "Latest *Homo Erectus* of Java: Potential Contemporaneity with *Homo Sapiens* in Southeast Asia," the Solo paper was published in *Science* on December 13, 1996.[2] As is typical of scientific papers, even this most dramatic of discoveries was couched in the most conservative, matter-of-fact language. The great majority of the paper was devoted to setting the context and describing the technical contortions that Henry and Jack put themselves through to ensure that the dates were correct, that nothing had been overlooked. Just two paragraphs at the end ventured comment on their significance.

For instance, we noted that, if proven correct, the new dates implied that "*Homo erectus* persisted much longer in Southeast Asia than anywhere else in the world." We also said that "if the Ngandong and Sambungmacan hominids represent a late-surviving sample of *H. erectus,* then the unilineal transformation in Southeast Asia from early and mid-Pleistocene *H. erectus* to early Australasian *H. sapiens*—with Ngandong and Sambungmacan as intermediate steps, as proposed by the multiregional theory for the origin of *H. sapiens*—is no longer chronologically possible." We ended the article with deliberate restraint: "The temporal and spatial overlap between *H. erectus* and *H. sapiens* in Southeast Asia, as implied by our study, is reminiscent of the overlap of the Neanderthals (*H. neanderthalensis*) and anatomically modern humans (*H. sapiens*) in Europe."

Each of these claims was noteworthy in itself. Combined, they demanded serious attention, which they duly received in the popular media. Every major newspaper in the world carried the story, but making the front page of the *New York Times,* above the fold, as it did on the day of publication in *Science,* was emblematic. "You can feign indifference to such things. But, you know what, they do matter," Susan says candidly. "Who wouldn't want their fifteen minutes of fame!" But the new Ngandong dates generated more than public notoriety, gratifying though that was. They also had a great impact on key issues relating to our recent evolutionary history.

Contemplating the possibility that *Homo erectus* might have survived until as late as 27,000 years ago is provocative. Although not exactly yesterday in terms of the time scales the modern human mind is used to, it is still shockingly recent for the survival of a species that is part of our evolutionary history and yet distinctly not us. Moreover, people like us—modern humans, *Homo sapiens*—lived at the same time as the Solo people, coexisting, if not side by side, then at least in much the same region of the world, just as Neanderthals and modern humans did in Europe, as we noted in the *Science* paper. We will return to this shortly, because the implications—in biological and philosophical realms—are profound.

What of the second of the three statements, that it was not chronologically possible for the Solo people to have been intermediate between earlier *Homo erectus* and the first settlers of Australia, as the supporters of multiregionalism had long claimed? If the Solo people lived until between 53,000 and 27,000 years ago, and the first Australians, modern humans as they were, landed on that island continent at least 60,000 years ago (based on archeological evidence; there are no fossil humans that old), then, logically, we must be correct in what we said in the paper—namely, that the Solo people cannot have been ancestors of people who lived *earlier* than they did, as much as 30,000 years earlier. Period. What had been a major pillar of the multiregional evolution hypothesis was swept away.

Not surprisingly, Thorne disagreed. In a flurry of e-mails that Thorne shot off just prior to publication, he said, "Even if this date is correct, which I think is extremely suspect, it has no impact at all on multiregionalism."[3] This was an extraordinary statement, given that having the Solo people temporally and anatomically intermediate between early *Homo erectus* in Java and the first Australians was central to multiregionalism. Remember Thorne and Wolpoff's statement on the matter in a popular British science magazine: "The most convincing evidence comes from Asia," they said, referring to the supposed sequence of ancestors from Sangiran through Solo to the modern Australians. Why, suddenly, did it make no difference that the Solo people weren't intermediates of any kind? Thorne gave several reasons; the principal one was the notion that Wolpoff had been vigorously promulgating—namely, that from a little less than 2 million years onward, only one human species existed, *Homo sapiens.*

Simply put, Thorne and Wolpoff's argument is that if, as the multiregional position holds, there is a continuous transformation of what is called *Homo erectus* to what is called *Homo sapiens,* then you cannot draw a clear line between the two at any single point in time and say that before this you have *Homo erectus* and after it *Homo sapiens;* you must lump them both into one species, *Homo sapiens* (because that species name was coined first). No matter that you have in the Solo people an anatomy that most anthropologists would call primitive, if not actually labeling it *Homo erectus,* and living at the same time as anatomically modern humans. To Thorne and Wolpoff, this is simply anatomical variation within a single species. By seeing it this way, say Thorne and Wolpoff, you can make all the problems of time, anatomy, and ancestry go away.

Of course they are right, in a sense. If all human individuals who lived between 2 million years ago and the present are members of *Homo sapiens,* then it doesn't matter that the primitive-looking Solo people coexisted with modern people: it just shows that you had a large degree of anatomical variation in Southeast Asia at that time. But we would ask, Does that have anything to do with evolution as it actually happens? We think not. If people were dealing with fossil antelopes rather than fossil humans, nobody would have any hesitation calling them different species. The current version of the multiregional model, a version in which dates don't matter anymore, is the single-species hypothesis in modern garb, and it is trying to avoid the fact that, although the multiregional evolution hypothesis says that all populations evolved toward *Homo sapiens,* here you obviously had people who were a relict population of *Homo erectus.*

As part of the regional continuity argument of the multiregional evolution hypothesis, Wolpoff and Thorne say that what you see in the anatomy of the first Australians is all Asian, with nothing African in them. "Then why do some of the early Australians look so much like Holocene [that is, recent] Africans?" asks Susan. She sees strong continuity of anatomy from early times through to the Solo people, and then when Australians appear on the scene there is a strong discontinuity, as modern people move into the region, bearing the mark of modern Africa, not ancient Java. "I've spent time with Milford [Wolpoff], and he gets all his casts out and we look at them, and we disagree. We just disagree." One of them has to be wrong, of course (or perhaps both).

Anthropologists, as we've seen, tend to have strong passions for what they consider correct, sometimes seeing evidence in different ways, other times not bothering to look at the evidence at all. "When the *Science* paper came out, I can't tell you how many times people said to me, 'Ngandong *can't* be that young because they're so primitive,'" recalls Susan. "These were people on my faculty. And I'd say, 'Have you read the *Science* paper? Have you looked at the dating work?' and they'd admit that they'd just seen it in the newspaper, not in *Science.* 'Go and read the paper,' I'd say. To which they'd reply, 'No one knows where the skulls come from anyway.' It's very frustrating."

We were certainly not trying to "curry favor with the Replacement people" with our paper, as Thorne contended, but it surely delivered unwelcome news to the multiregionalists. As the Berkeley anthropologist Clark Howell remarked at the time of publication, "If it's true, it's another knock against the multiregional hypothesis."[4] Ian Tattersall of the American

Museum of Natural History was more blunt: "Multiregionalism is dead."[5]

Three's Company, and More

The picture of human evolution used to be simple. The human family was born in Africa, to which continent it was confined for a very long time. Then, about a million years ago, bands of a relatively advanced form of human—*Homo erectus*—migrated into eastern Asia. All later forms of human were descendants of this species, the product of a steady modernization process that included the enlargement of the brain and an anatomy that was becoming less robust. Before modern humans—that is, *Homo sapiens*—arrived on the scene, there were populations of people who were intermediate between classic *Homo erectus* and *Homo sapiens,* lumped under the collective term "archaic *sapiens.*" These were considered to be penultimate states in the evolution of modern humans.

The first part of the story remains true: the human family was born in Africa, about 5 or 6 million years ago. But we know from our work on the Mojokerto child and Sangiran that the second event—the migration out of Africa—occurred much earlier than was thought, close to 2 million years ago. So that bit of the story, the 1 million years, is wrong. And our recent redating of the Solo people has contributed to a growing understanding that the rest of the story is wrong, too. Had anthropologists paid more attention to basic biological and evolutionary patterns seen in other animal species, they might more quickly have recognized the shortcomings of the old scenario. The long, slow transformation of *Homo erectus* into *Homo sapiens* embodies the notion of progressive evolution in a particular direction, the steady improvement of a particular attribute—in this case brain power and size. This beguiling image used to be popular, but modern evolutionary biologists know it to be untrue. Species, once they have evolved, tend to remain much the same for long periods of time. And change, when it comes, tends to happen quickly, in punctuational events. Also, it is very unusual in the rest of the animal world to have just two species, one turning into the other, over this stretch of time. Typically, multiple species will come and go, so that at any one time several will coexist. This is the image toward which some anthropologists are moving, and to which the new date for Ngandong contributes. But there is resistance.

There has been a long tradition in anthropology to view anatomical differences among specimens simply as variation within the species, not differences between species. This was the assessment of the first Neanderthal specimen in the mid-nineteenth century: it was judged to be a primitive race of modern people. Only the Irish anatomist William King was bold enough to suggest that it was something other than *Homo sapiens,* that is, *Homo neanderthalensis.* It is true that between the 1930s and 1950s there was a burst of enthusiasm for so-called splitting—that is, seeing any small difference as a signal of uniqueness—and for a time every human fossil that came out of the ground was baptized with its own species name (and, often, its own genus name as well). The Harvard evolutionary biologist Ernst Mayr put an end to this nomenclatural excess in 1950, when he consolidated many to a few species and even argued that they should all be under the genus name *Homo,* with no *Australopithecus* at all. When anthropologists embraced Mayr's suggestion about consolidation (although not the part about having just one genus), they were, as Ian Tattersall has put it, "reverting to a longstanding mind set."[6] Anthropologists these days tend to be lumpers, not splitters.

That mind-set's hold, although still strong, is now diminishing, for which there are several reasons, particularly when we are considering human prehistory of, say, the past half-million years. One is practical, the other is more philosophical or sociological.

The practical issue, which is faced by paleontologists of all stripes, not just those who deal with human fossils, is how to distinguish one species from another based on differences in lumps and bumps on a few petrified bones. After all, lumps and bumps are all paleontologists have to go on. Even biologists who study living creatures fuss over what a species really is, with terms such as "biological species concept," "recognition species concept," and "phylogenetic species concept" reflecting different philosophical positions. The first, with which most people have at least a passing familiarity, is that males and females of different species don't mate; that is, they are reproductively isolated from each other. This works fine with, say, a horse and a chicken, but closely related species do in fact mate, and sometimes they even produce viable offspring. Whatever its merits or demerits, the biological species concept is of little use to paleontologists, since fossil bones don't mate.

The recognition species concept looks at behaviors or markings that allow males and females to recognize each other as potential mates while leaving them unmoved by other species whose behavior or markings don't declare, "Let's reproduce together." Again, however, and for the same reason as with the first concept, the paleontologist can't do much with this one. The phylogenetic species concept is more in the paleontologist's territory, because it identifies species in the context of their evolution. But you still need to identify a species as a species, and you need to be able to do that from lumps and bumps on bones. To illustrate how very difficult this can be, Alan Walker likes to say that if you went to the coastal forests of Kenya and shook the monkeys out of the trees, stripped them of flesh and other soft tissues, and then examined the bones for diagnostic features, you would conclude that there was just one species, while in reality you would know that there were at least a dozen, the specific differences being embodied in the markings and behaviors. This tale is salutary for anthropologists because it implies that if they are erring in their estimates of how many human species existed at any one time, they are probably erring in the direction of an underestimate rather than an overestimate.

What message does this have for how the later period of human prehistory is interpreted? It suggests that unless humans are different from all other animals (for some unexplained reason), several species coexisted on Earth, perhaps sometimes in the same geographical region, until a little less than 30,000 years ago. Look at the fossil record for this time, and you see that among the human skulls that have been found in Africa, Asia, and Europe there are a lot of anatomical differences. Is this evidence of three or four, or even more, human species? Until recently, textbooks lumped them all together under the misleading term "archaic *sapiens,*" which was meant to imply that these separate populations around the world were in the final stages of their evolutionary journey to being fully modern humans. This is highly unlikely, for no other reason than that it does not conform to the pattern of evolutionary change that happens elsewhere in the world. As we said earlier, if these specimens were fossil antelopes, biologists would happily describe the different fossil populations as different species. The term "archaic *sapiens*" is misleading, because it implies that the humans in question were slightly primitive versions of us, whereas in fact they were fully established versions of themselves, and were not *Homo sapiens.*

But the lumping mold is beginning to crack, beginning with the now widely accepted proposal that the Neanderthals were a separate species, *Homo neanderthalensis,* as King suggested a century ago. With the notion that recognizing different human species in relatively recent times is no longer politically incorrect, other species are being proposed, such as *Homo heidelbergensis* and *Homo helmei.* The details of which fossils are proposed as being these species, and why, don't concern us here (see the diagram on page 144). More important is the growing recognition among some professional anthropologists that humans are much the same as other animals in our pattern of evolution, that we were not propelled to our present state by some unique force of nature, even though we may be something special in the world now.

Why did most anthropologists insist on treating anatomical differences among recent fossil humans in ways that they would not if the fossils were antelopes or, for that matter, if the fossils were early humans? Tattersall has suggested that brain size obscured all other anatomical features in anthropologists' assessment of the so-called archaic *sapiens* fossils. If the fossil humans had big brains, they must be us, or at least very nearly us, was the line of thinking. It was as if anthropologists, on seeing a big brain, could not bring themselves to see anything other than *"sapiens,"* even though other parts of the cranial anatomy were very different from modern people's. Like deer caught in car headlights, scientists can freeze in one intellectual position and be blind to everything else.

The notion of a steady, progressive sapiensization of *Homo* from 2 million years ago to the recent past was comforting in its implied inevitability, the inevitability that we would arrive one day. We saw in an earlier chapter that this was very much how anthropologists thought of evolution in the early decades of this century, and we can see how persistent the core of that belief has been. If, as the Out of Africa hypothesis suggests, modern humans evolved in an isolated population around 150,000 years ago, with its descendants subsequently moving into the rest of the world, then our arrival as modern humans has a much more chancy, uncertain feel to it. Why that population and not another? Why any population at all? Chance events play a large part in evolutionary history, biologists have come to realize. There was no inevitability to the evolution of *Homo sapiens.*

If what triggered that event—some small change in the local ecosystem, some chance mutation that conferred advantageous properties, such as fully modern language—had not occurred, the world might still be populated by multiple human species. Neanderthals might still be here, and *heidelbergensis* and *helmei,* and whatever other species existed at that time but have yet to be identified. And perhaps even Solo people, too. But they are not here, and most likely that is because the world was not a big enough place for them and for modern humans, ourselves, when we did evolve. Their disappearance might simply have been the consequence of a far greater efficiency of modern humans in exploiting natural resources, as archeological evidence indicates was true. These other, now extinct, human species simply lost out in economic competition, losers in a game of who can survive best in a world of limited resources.

Or the incoming population of modern humans might have hastened their extinction in ways that we know are all too human, that of violence. There is absolutely no evidence in the prehistoric record that this was what happened, but we cannot escape it as a possibility. Our having caused the extinction of species of humans very much like ourselves through economic competition is regrettable enough, but our having brought about extinction through violence is hard to contemplate. Of course, if all humans at that time in human prehistory were members of archaic *sapiens,* then the possibility that we are the descendants of a population of people who pushed to extinction a handful of species that were very much like us would simply disappear: there would have been no handful of species to push into extinction, just one big variable species. If all human populations in the world were *sapiens,* then the arrival of modern humans would have brought about local extinctions of *sapiens* folk, not complete extinction of individual species.

It's a happier way to imagine the past. But new thinking in anthropology, propelled in part by discoveries such as the young age of the Solo people, no longer leaves tenable that anodyne alibi. Almost certainly, everyone in the world today is a descendant of a population of newly emerged humans who, in their inexorable spread across the world, caused the extinction of species of humans that were like us, but weren't us. We wonder how much time must pass before that, too, becomes a politically correct version of our history, like it or not.

Notes

Chapter 1: Tales a Child Can Tell

1. G. H. Curtis, "Establishing a relevant time scale in anthropological and archeological research," *Philosophical Transactions of the Royal Society,* B, vol. 292 (1981), pp. 2—20, p. 16.
2. A. Deino et al., "$^{40}Ar/^{39}Ar$ dating in paleoanthropology and archeology," *Evolutionary Anthropology,* vol. 6 (1998), pp. 63—75.

Chapter 2: The Road to Trinil

1. This quotation, along with all others from Wallace in this chapter, is taken from A. R. Wallace, "Alfred Russel Wallace declares Java 'The finest tropical island in the world,'" 1869, in J. R. Rush, *Java: A Travellers' Anthology* (Oxford University Press, 1996), pp. 67—83.
2. H. Movius, "Early man and Pleistocene stratigraphy in Southern and Eastern Asia," Papers of the Peabody Museum of American Archeology and Ethnology, Harvard University, vol. XIX, no. 3 (1944), pp. 70—108; H. de Terra, "Pleistocene geology and early man in Java," *Transactions of the American Philosophical Society,* n.s. 32, part 5 (1943), pp. 437—466.

3. G. H. R von Koenigswald, *Meeting Prehistoric Man* (Thames and Hudson, 1956), p. 22.

Chapter 3: On to Mojokerto

1. G. H. R von Koenigswald, *Meeting Prehistoric Man* (Thames and Hudson, 1956), p. 80.
2. D. Ninkovich et al., "Paleographic and geologic setting for early man in Java," in R. A. Scrutton and M. Talwani, *The Ocean Floor* (John Wiley & Sons Ltd., 1982), pp. 211–227.

Chapter 4: The Lure of the Missing Link

1. G. H. R. von Koenigswald, *Meeting Prehistoric Man* (Thames and Hudson, 1956), p. 25.
2. Cited in B. Theunissen, *Eugène Dubois and the Ape-Man from Java* (Kluwer Academic Publishers, 1989), p. 23.
3. Ibid., p. 23.
4. Ibid., p. 48.
5. Ibid., p. 25.
6. E. Haeckel, *The History of Creation,* vol. 2 (D. Appleton and Company, 1876), pp. 293–294.
7. T. H. Huxley, *Evidence as to Man's Place in Nature* (D. Appleton and Company, 1892), p. 150.
8. Cited in Theunissen, *Eugène Dubois,* p. 33.
9. Ibid., p. 49.
10. Ibid., p. 13.
11. Ibid., p. 31.
12. Cited in von Koenigswald, *Meeting Prehistoric Man,* p. 28.
13. Ibid., p. 27.
14. Ibid., p. 27.

Chapter 5: Dubois's Story: Link No Longer Missing

1. Cited in B. Theunissen, *Eugène Dubois and the Ape-Man from Java* (Kluwer Academic Publishers, 1989), p. 39.
2. All quotations in this paragraph are from Theunissen, *Eugène*

Dubois, p. 40.
3. Ibid., p. 54.
4. Ibid., p. 55.
5. G. H. R. von Koenigswald, *Meeting Prehistoric Man* (Thames and Hudson, 1986), p. 25.
6. Cited in Theunissen, *Eugène Dubois,* p. 57.
7. Ibid., p. 58.
8. Ibid., p. 58.
9. Ibid., p. 58.
10. Ibid., p. 73.
11. Ibid., p. 65.
12. Von Koenigswald, *Meeting Prehistoric Man,* p. 25.
13. Cited in Theunissen, *Eugène Dubois,* p. 115.
14. Ibid., p. 112.
15. Ibid., p. 121.
16. Ibid., p. 169.
17. Both quotations in this paragraph are from Theunissen, *Eugène Dubois,* p. 152.
18. Ibid., p. 156.
19. Ibid., p. 154.
20. Von Koenigswald, *Meeting Prehistoric Man,* p. 37.
21. Cited in Theunissen, *Eugène Dubois,* p. 156.
22. Quotations in this paragraph are from von Koenigswald, *Meeting Prehistoric Man,* pp. 81—82.
23. Ibid., p. 95.
24. Ibid., p. 97.
25. Both quotations are from von Koenigswald, *Meeting Prehistoric Man,* p. 98.
26. Ibid., p. 99.
27. Cited in Theunissen, *Eugène Dubois,* p. 159.
28. Cited in Theunissen, *Eugène Dubois,* p. 163.
29. Cited in Theunissen, *Eugène Dubois,* p. 165.

Chapter 6: The Child Has a Date

1. C. C. Swisher et al., "Age of the earliest known hominids in Java, Indonesia," *Science,* 25 February 1994, pp. 1118—1121.

2. M. W. Browne, "Asian fossil prompts new ideas on evolution," *New York Times,* 24 February 1994, p. A1.
3. C. Petit, "New fossil date shakes up ideas on evolution," *San Francisco Chronicle,* 24 February 1994, p. 1.
4. M. D. Lemonick, "How man began," *Time,* 14 March 1994, pp. 80—87; quotation from p. 82.
5. R. Lewin, "Human origins: the challenge of Java's skulls," *New Scientist,* 7 May 1994, pp. 36—40; quotation from pp. 36, 37.
6. Ibid., p. 36.
7. A. Gibbons, "Rewriting—and redating—prehistory," *Science,* 25 February 1994, pp. 1087—1088.
8. H. de Terra, "Pleistocene geology and early man in Java," *Transaction of the American Philosophical Society,* n.s. 32, part 5 (1943), pp. 437—464; quotation from p. 442.
9. Ibid., p. 443.
10. H. L. Movius, "Early man and Pleistocene stratigraphy in southern and eastern Asia," Papers of the Peabody Museum of American Archaeology and Ethnology, Harvard University, vol. XIX, no. 3 (1994), p. 81.
11. N. Toth el al., "Pan the tool-maker: investigations into the stone tool-making and tool using capabilities of a Bonobo," *Journal of Archeological Research,* vol. 20, (1993), pp. 81—91.

Chapter 7: Rocky Marriage, Painful Separation

1. T. D. White, cited in R. Lewin, *Bones of Contention,* 2d ed. (University of Chicago Press, 1997), p. 272.
2. Interview with the authors, San Francisco, November 11, 1997.
3. D. C. Johanson and T. D. White, "A systematic reassessment of early African hominids," *Science,* vol. 203 (1979), pp. 321—330.
4. See, for instance, Lewin, *Bones of Contention.*
5. D. C. Johanson and J. Shreeve, *Lucy's Child* (William Morrow, 1989), p. 28.
6. Interview with the authors, Berkeley, CA, November 12, 1997.
7. P. Renne, "Institute of Human Origins Breakup," *Science,* vol. 265 (1994), pp. 721—722.
8. C. C. Swisher and D. R. Prothero, "Single-crystal $^{40}Ar/^{39}Ar$

dating of the Eocene-Oligocene transition in North America," *Science*, vol. 249 (1990), pp. 760—762.
9. R. Lewin, "Rock of Ages: cleft by laser," *New Scientist*, September 26, 1991, pp. 36—40.
10. P. Renne, memorandum to IHO Science Committee, "IHO-GC role in Ar/AR dating history," April 22, 1994.
11. Plaintiff's memorandum of points and authorities, June, 8, 1994, p. 3.
12. Ibid., p. 1.
13. Ibid., p. 9.
14. Supplement to memorandum of points and authorities in opposition to request for temporary restraining order, June, 10, 1994, pp. 1—2.
15. Cited in the *Wall Street Journal*, June, 2, 1995.

Chapter 8: On the Cusp of Humanity

1. A. Keith, *A New Theory of Human Evolution* (Philosophical Library, 1949), p. 161.
2. G. Elliot Smith, *Essays on the Evolution of Man* (Oxford University Press, 1924), p. 40.
3. M. Cartmill, "Four legs good, two legs bad," *Natural History*, November 1983, pp. 65—78; quotation from p. 68.
4. Elliot Smith, *Essays*, p. 79.
5. J. Stern and R. Susman, "The locomotor behavior of *Australopithecus afarensis*," *American Journal of Physical Anthropology*, vol. 60 (1983), pp. 279—317; quotation from p. 280.
6. W. K. Gregory, "Two views of the origin of man," *Science*, May, 17, 1927, pp. 601—605; quotation from p. 602.
7. M. Goodman, "Molecular anthropology," *McGraw-Hill Encyclopedia of Science and Technology* (McGraw-Hill, 1999).
8. M. Goodman et al., "Toward a phylogenetic classification of primates based on DNA evidence complemented by fossil evidence," *Molecular Phylogenetics and Evolution*, vol. 9 (1998), pp. 585—598.
9. C. Darwin, *The Descent of Man, and Selection in Relation to Sex* (Princeton University Press, 1981), p. 142.

10. Ibid., p. 141.
11. Ibid., p. 144.
12. P. S. Rodman and H. M. McHenry, "Bioenergetics of hominid bipedalism," *American Journal of Physical Anthropology,* vol. 52 (1980); quotation from p. 106.
13. L. A. Isbell and T. P. Young, "The evolution of bipedalism in hominids and reduced group size in chimpanzees," *Journal of Human Evolution,* vol. 30 (1996), pp. 389–397.

Chapter 9: A Change of Body

1. A. Walker and P. Shipman, *The Wisdom of the Bones* (Alfred A. Knopf, 1996).
2. H. M. McHenry, "Behavioral ecological implications of early body size," *Journal of Human Evolution,* vol. 27 (1994), pp. 77–87.
3. H. M. McHenry, "Sexual dimorphism in fossil hominids and its sociological implications," in J. Steele and S. Shennan, eds., *The Archeology of Human Ancestry* (Routledge, 1996), pp. 91–109.
4. A. Walker and P. Shipman, *Wisdom of the Bones,* p. 168.
5. P. Shipman and A. Walker, "The costs of becoming a predator," *Journal of Human Evolution,* vol. 18 (1989), pp. 373–392.
6. S. Washburn and C. S. Lancaster, "The evolution of hunting," in R. B. Lee and I. DeVore, eds., *Man the Hunter* (Aldine Publishing Company, 1968), pp. 293–303; quotation from p. 293.
7. R. Ardrey, *The Hunting Hypothesis* (Collins, 1976).
8. R. Potts, cited in R. Leakey and R. Lewin, *Origins Reconsidered* (Doubleday, 1992), p. 182.
9. P. Colinvaux, *Why Big, Fierce Animals Are Rare* (Princeton University Press, 1978).

Chapter 10: A Change of Mind

1. R. Susman, "Hand function and tool behavior in early hominids," *Journal of Human Evolution,* vol. 35 (1998), pp. 23–46.
2. T. Wynn and W. C. McGrew, "An ape's view of the Oldowan," *Man,* n.s., vol. 24 (1989), pp. 383–398; quotation from p. 388.

3. See, for example, D. Bickerton, *Language and Human Behavior* (University of Washington Press); T. W. Deacon, *The Symbolic Species: The Co-evolution of Language and the Brain* (W. W. Norton, 1997); R. Dubar, *Grooming, Gossip, and the Evolution of Language* (Harvard University Press, 1998); P. Lieberman, *Eve Spoke* (W. W. Norton, 1998); S. Mithen, *The Prehistory of the Mind,* (Thames and Hudson, 1996); S. Pinker, *The Language Instinct* (William Morrow, 1994).
4. D. Premack, cited in Pinker, *The Language Instinct,* p. 367.
5. Pinker, *The Language Instinct,* p. 18.
6. Ibid., p. 24.
7. R. Dunbar, "The social brain hypothesis," *Evolutionary Anthropology,* vol. 6 (1998), pp. 178—190.
8. Lieberman, *Eve Spoke.*
9. W. Noble and I. Davidson, *Human Evolution, Language, and Mind* (Cambridge University Press, 1996).
10. R. White, "Rethinking the Middle/Upper Paleolithic transition," *Current Anthropology,* vol. 23 (1982), pp. 169—189.
11. D. Falk, "3.5 million years of hominid brain evolution," *Seminars in Neuroscience,* vol. 3 (1991), pp. 409—416; quotation from p. 410.

Chapter 11: The Origin of Modern Humans

1. L. Brace, "The fate of the classic Neanderthals," *Current Anthropology,* vol. 5 (1964), pp. 3—43.
2. R. L. Cann et al., "Mitochondrial DNA evolution and human evolution," *Nature,* vol. 325 (1987), pp. 31—36; quotation from p. 35.
3. F. J. Ayala, "The myth of Eve: molecular biology and human origins," *Science,* vol. 270 (1995), pp. 1930—1936; quotation from p. 1930.
4. "The Search for Adam and Eve," *Newsweek,* January 11, 1988, pp. 46—52; quotation from p. 46.
5. J. Shreeve, "Argument over a woman," *Discover,* Aug. 1990, pp. 52—59; quotation from p. 52.
6. Ibid., p. 57.

7. Ibid.
8. E. Zubrow, "The demographic modelling of Neanderthal extinction," in P. Mellars and C. B. Stringer, eds., *The Human Revolution* (Princeton University Press, 1989), pp. 212–231; quotation from p. 229.
9. J. N. Wilford, "Critics batter proof of an African Eve," *New York Times,* May 19, 1992, C1.
10. H. M. Watzman, "Question of human origins debated as scientists put to rest the idea of a common African ancestor," *Chronicle of Higher Education,* September 16, 1992, p. A7.
11. Quoted in A. Gibbons, "Mitochondrial Eve refuses to die," *Science,* vol. 259 (1993), pp. 1249–1250; quotation from p. 1249.
12. A. Gibbons, "Y chromosome shows that Adam was an African," *Science,* vol. 278 (1997), pp. 804–805.
13. M. Ruvollo, *Molecular Phylogenetics and Evolution,* vol. 5 (1996), pp. 202–219; quotation from p. 202.
14. C. B. Stringer, cited in P. Kahn and A. Gibbons, "DNA from an extinct human," *Science,* vol. 277 (1997), pp. 176–178; quotation from p. 176.
15. M. Krings et al., "Neanderthal DNA sequences and the origin of modern humans," *Cell,* vol. 90 (1997), pp. 19–30; quotation from p. 19.
16. L. Cavalli-Sforza, quoted in Krings et al., p. 177.
17. A. S. Brooks et al., "Dating and context of three Middle Stone Age sites with bone points in the Upper Semliki Valley, Zaire," *Science,* vol. 268 (1995), pp. 548–556; S. McBrearty et al., "Variability in traces of Middle Pleistocene hominid behavior in the Kapthurin Formation, Baringo, Kenya," *Journal of Human Evolution,* vol. 30 (1996), pp. 563–580.
18. M. Wolpoff and A. Thorne, "The case against Eve," *New Scientist,* June 22, 1991, pp. 37–41; quotation from p. 39.
19. Ibid., p. 39.
20. C. B. Stringer, cited in J. Shreeve, "*Erectus* rising," *Discover,* September 1994, pp. 80–89; quotation from p. 89.
21. M. H. Wolpoff, interview with the author, April 1994.

Chapter 12: Headhunters at Ngandong

1. G. H. R. von Koenigswald, *Meeting Prehistoric Man* (Thames and Hudson, 1956), p. 66.
2. Ibid., p. 68.
3. Ibid., p. 65.
4. H. F. Osborn, "Is the ape-man a myth?" *Human Biology,* vol. 1 (January 1929), pp. 4–9; quotation from p. 4.
5. Ibid., p. 70.
6. Von Koenigswald, *Meeting Prehistoric Man,* p. 74.
7. Ibid., p. 74.
8. Ibid., p. 72.
9. Ibid., p. 75.
10. Ibid, pp. 78–79.
11. G. H. R. von Koenigswald, cited in P. V. Tobias, *The life and work of Dr. G. H. R. von Koenigswald* (Verlag Waldemar Kramer, 1984), pp. 58–59.
12. Ibid.
13. G. H. R. von Koenigswald, Introduction to F. Weidenreich, "Morphology of Solo Man," *Anthropological Papers of the American Museum of Natural History,* vol. 43 (1951), part 3, pp. 204–289; quotation from p. 213.
14. H. Shapiro, Foreword to above, p. 205.
15. M. Day, *Guide to Fossil Man,* 4th ed. (University of Chicago Press, 1986).
16. Von Koenigswald, *Meeting Prehistoric Man,* p. 70.
17. F. Weidenreich, "The Keilor skull: A Wadjak skull from southeast Asia," *American Journal of Physical Anthropology,* vol. 3 (1945), pp. 21–33; quotation from p. 31.
18. A. P. Santa Luca, "The Ngandong fossil hominids," *Yale University Publications in Anthropology,* no. 78 (1980).
19. M. Wolpoff and A. Thorne, "The case against Eve," *New Scientist,* June 22, 1991, pp. 37–41; quotation from p. 39.
20. Ibid.

Chapter 13: Facing the Inescapable

1. G. H. R. von Koenigswald, Introduction to F. Weidenreich, "Morphology of Solo Man," *Anthropological Papers of the American Museum of Natural History,* vol. 43 (1951), part 3, pp. 205–289; quotation from p. 218.
2. C. C. Swisher et al., *Science,* vol. 274 (1996), pp. 1870–1874.
3. A. Thorne, e-mail to C. Petit, December 11, 1996.
4. Interview with the author, December 1996.
5. Interview with the author, December 1996.
6. I. Tattersall, "Species concepts and species identification in human evolution," *Journal of Human Evolution,* vol. 22 (1992), pp. 341–349; quotation from p. 346.

Index

Index numbers in italics refer to illustrations.